Instabilities in MOS Devices

Electrocomponent Science Monographs

A series edited by D.S. Campbell

Volume 1 *Instabilities in MOS Devices*, by J. R. Davis

Other volumes in active preparation include:

Capacitors, by D. S. Campbell

Dielectric Films on Gallium Arsenide, by W. F. Croydon and E. H. C. Parker

Ferroelectric Transducers, by J. M. Herbert

Electronic Component Quality Assurance and Reliability, by J. Møltoft

Instabilities in MOS DEVICES

J. R. Davis

Post Office Research Centre
Ipswich, England

GORDON AND BREACH SCIENCE PUBLISHERS

New York London Paris

Gordon and Breach, Science Publishers, Inc.
One Park Avenue
New York, NY10016

Gordon and Breach Science Publishers Ltd.
42 William IV Street
London WC2N 4DE, England

Gordon & Breach
7–9 rue Emile Dubois
F-75014 Paris, France

Library of Congress Cataloging in Publication Data

Davis, John Richard, 1951—
Instabilities in MOS devices.

(Electrocomponent science monographs; v. 1)
Bibliography: p.
Includes index
1. Metal oxide semiconductors. I. Title.
II. Series.
TK7871.99.M44D38 621.3815'2 80-25156
ISBN 0-677-05590-0

Library of Congress Catalog Card Number: 80-25156. International Standard Book Number: 0 677 05590 0. Printed in the United Kingdom.

To My Parents

Introduction to the Series

This series of monographs, in the field of electronic and electrical component science and technology, publishes authoritative reviews in the subjects, written by acknowledged experts. Reviews of this type are normally collected in volumes containing several chapters on different, relatively unrelated subjects and it is not usual for subscribers to want all of the reports which are bound into the one volume. This situation limits the availability of each review due to the cost of the complete volume. This present series is aimed at solving this problem by publishing each review as a separate monograph.

As well as volumes concerned with particular components, the monograph series will also include reviews on associated subsystems, particularly with regard to the development and use of LSI, VLSI and hybrid technology, etc.

The publications are designed for use by final-year undergraduates, postgraduates, and engineers working in the field. It is hoped that the books will prove stimulating and useful.

D.S. Campbell

LOUGHBOROUGH UNIVERSITY OF TECHNOLOGY

Contents

Page

Preface

This monograph is concerned with the instabilities which cause the parameters of MOS devices to differ from the values predicted by first-order theory and, more importantly, which result in drifts when the devices are exposed to temperature, voltage, radiation or other stresses. It is intended as an introductory text for undergraduate and graduate students of solid-state physics or electronics, and as a concise review for semiconductor processing, reliability and quality assurance engineers. A basic knowledge of solid-state physics, integrated circuit technology and semiconductor device operation (such as is given in "Introduction to Solid-State Physics" by Kittel (Wiley, 1966), "MOS Integrated Circuits", edited by Penney and Lau (Van Nostrand) and "Physics of Semiconductor Devices" by Sze (Wiley, 1969), respectively) is assumed.

The text was originally written in 1977 as an MSc Critical Survey, and has been rewritten to include additional material which has been published since then. In a text of this length it is not possible to give a complete survey of all the available literature on MOS devices. Instead, the classic papers on each topic are cited, as well as a careful selection of the most recent publications, thus giving a general account of the present knowledge in a given area, and allowing researchers interested in a particular aspect of any topic to locate the full literature.

The term "MOS device" is not used in its strictly correct sense in this text, but is taken to include all the devices in the MOS sections of the semiconductor manufacturers' catalogues, such as those employing polysilicon gates and insulators other than simple oxides (such as silicon nitride), and which should properly be referred to as "IGFETs". Also, the text deals exclusively with silicon since a large fraction of the literature deals with it, and because of its present nearly universal use in integrated circuits (notwithstanding the rapidly growing interest in the compound semiconductors).

At this stage it is worthwhile pointing out that the references quoted can broadly be regarded as coming from three categories of sources, so that the importance and relevance of any paper should be assessed bearing in mind the author's background. Firstly, the MOS system is used by research workers (principally at universities) as a tool for investigating the physics of surfaces and interfaces (the silicon/silicon dioxide interface is perhaps the best understood of all solid/solid interfaces, although large gaps in the available knowledge remain). These workers are not influenced by any commercial pressures, but unfortunately they do not always have access to state-of-the-art processing technologies, so that their findings do not always reflect directly on the instabilities observable in real integrated circuits.

The second group of workers are employed by the semiconductor processing houses, and thus are liable to underplay any instability. In the early days of MOS technology these workers published a large amount of the available information, reporting on the gross instabilities of the time. With the advent of marketable MOS devices, however, they were naturally reluctant to report on the instabilities in their company's technology (except with hindsight), and concentrated instead on attempting to prove the reliability of their products under less arduous stressing conditions. Also, the fine details of the technology required to produce stable devices with high yields have become valuable commercial secrets, and are not freely available.

The third category of researchers work for organisations which are large users of MOS devices, and hence are concerned to detect any potential sources of unreliability. These workers are often sponsored, directly or indirectly, by governments, nationalized industries, telecommunication organisations, space agencies, and by the military. Perhaps the most authoritative sources of information, combining the facilities to produce state-of-the-art devices with the freedom to report objectively on them, are those organisations which have advanced processes for in-house use, or for comparison with those available commercially.

The text is organised as follows. Chapter 1 gives a brief review of MOS and field-effect transistor theory, and considers the structure of silicon dioxide and its interface with silicon. Chapter 2 describes the experimental techniques most commonly used to explore the electrical properties of the MOS system, and then Chapters 3 to 7 each deal with a specific instability mechanism, in approximately the chronological order in which they have limited the performance of practical devices. Conduction and dielectric breakdown problems, whilst not being true instabilities, are becoming ever more important to the reliability of MOS devices, and are dealt with in Chapter 8. Finally, the list of references includes a table of papers arranged according to their topics.

It is a pleasure to acknowledge the help that I have received from my colleagues at the British Post Office Research Centre, Martlesham, in writing this book. In particular, I should like to thank Mr P Mellor and Dr R Enoch for their overall encouragement and constructive criticism; Dr F Reynolds, Mr G Procter and Drs S O'Hara and A Elliott for their reports from certain conferences, and the latter also for discussions on the behaviour of surface states; Mr P Dunn for discussions on various aspects of device processing; Mr A Watt for access to unpublished results on some specific instability mechanisms, and Mr D Kennett for proof-reading parts of the text.

In addition, the references cited could not have been gathered without the assistance of the staff of the Research Centre Library, and many drawings were prepared with the help of the Photographic Group. Finally, I am indebted to Miss S White for typing the entire manuscript.

J R Davis
June 1980

1

Introduction

1.1 Historical

The possibility of utilising the surface field effect to produce a solid-state amplifier was originally proposed as long ago as the 1930s by Lilienfeld [1] and Heil [2] and demonstrated in a crude form in the 1940s by Shockley and Pearson [3]. At that time, however, there was insufficient knowledge of semiconductor surfaces, or ability to control their characteristics, to produce a practical device so, with the invention of the junction (bipolar) transistor at about the same time, the Insulated Gate Field Effect Transistor (IGFET) was not exploited further for many years. Following the development of silicon planar technology for high-performance bipolar transistors, Kahng and Attala [4] fabricated the first modern IGFET in 1960 by using a thermally oxidized silicon structure. Although the importance of the new device, with its very high input impedance, small size and simple processing, was immediately apparent, early samples were far too unstable to be of any commercial value; indeed, their characteristics were degraded simply by displaying them on a curve tracer. With the growing importance of digital systems, however, for which the MOS transistor is particularly suited, a large amount of research effort was expended on improving the control over the silicon/silicon dioxide interface, so that by 1967 Schlegel [5] was able to compile a bibliography containing over 700 references. MOS transistors and integrated circuits became commercially viable at this time, but they were not reliable enough to be considered for use in equipment requiring long life-times, such as telecommunication systems.

Many further improvements have been made since 1967. An excellent bibliography by Agajanian [6] in 1977 listed

560 references on silicon dioxide (categorized under various topics) published in the previous decade. Because of its near-ideal properties, the Si/SiO_2 interface has attracted attention from physicists as well as semiconductor engineers, to such an extent that the topic now commands its own conference [7].

Semiconductor manufacturers sometimes claim that the instabilities of the MOS system have been completely overcome. Certainly, the grosser instabilities are now understood to a degree which allows them to be removed, or at least controlled, by careful processing. However, limitation of one instability often results in the uncovering of a second, more subtle one, so that with the ever increasing requirement for highly stable devices with characteristics closely approaching the ideal, theoretically predicted case, there is a continued need to investigate the MOS system. This is especially so, since the more complicated processing and design technologies now being introduced to improve circuit performance result in increasing scope for new instability mechanisms. With this in mind, this book will attempt to describe the present understanding of the various instability mechanisms, together with the techniques available for investigating them, and their implications for long-term device reliability.

1.2 Basic MOS Theory

It is not planned to describe the detailed theory of the MOS system in this section, but a resume of the basic characteristics of the ideal device will be given, together with some indication of the variations found in practical cases. For a more extensive treatment, the reader is referred to the excellent book by Sze [8], or to Many, Goldstein and Grover [9]. Many practical measurements, as well as the basic theory of the system, are presented in a classic paper by Nicollian and Goetzberger [10], the latter author also having published sets of ideal curves of the MOS system [11], which are very useful for comparison purposes.

The band diagram for an ideal MIS capacitor structure is shown in Figure 1.1. In this ideal case, with no bias applied to the metal electrode, there is no energy difference between the metal work function ϕ_M and the semiconductor work function ϕ_S given by

$$\phi_S = \chi + E_G/2q + \psi_B \tag{1.1}$$

where χ is the semiconductor electron affinity, E_G is the

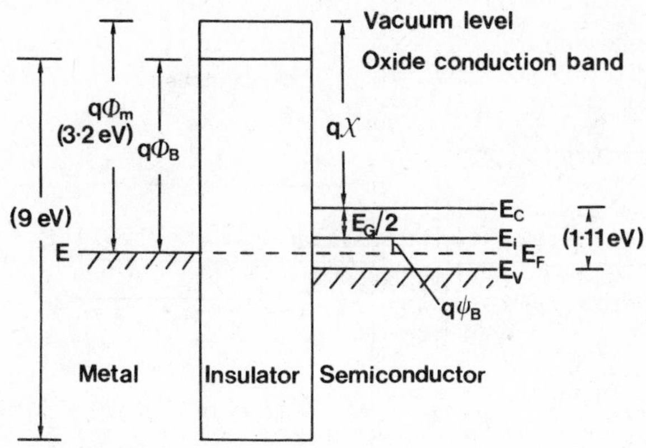

Figure 1.1 Idealized band diagram of the Metal-Insulator-Semiconductor system with zero applied voltage and with no work-function difference. Values in brackets refer to the Al-SiO_2-Si structure.

bandgap and ψ_B is the potential difference between the Fermi level E_F and the intrinsic Fermi level E_i. In this case, the bands are flat right up to the semiconductor surface (the 'flat-band' condition) for zero applied voltage. We are interested in modifying the surface space-charge region by means of altering the surface potential ψ_s; this can be achieved by applying a voltage to the gate electrode. For a p-type semiconductor, the three important states are shown in Figure 1.2; the case for n-type semiconductor simply involves reversing the sign of the applied voltage. We are particularly interested in the inversion condition, where the mobile carriers at the surface (electrons) are of the opposite type to those in the bulk (holes) and the conducting surface layer is separated from the bulk by a depletion region. The onset of strong inversion (when the space charge in the surface region is equal and opposite to that of the bulk) can be shown to occur at

$$\psi_s \text{ (inversion)} \simeq 2\psi_B \simeq \frac{2kT}{q}.\ln\left(\frac{N_A}{n_i}\right) \qquad (1.2)$$

where the symbols have their normal meanings. The relationship between the voltage on the field plate and the surface

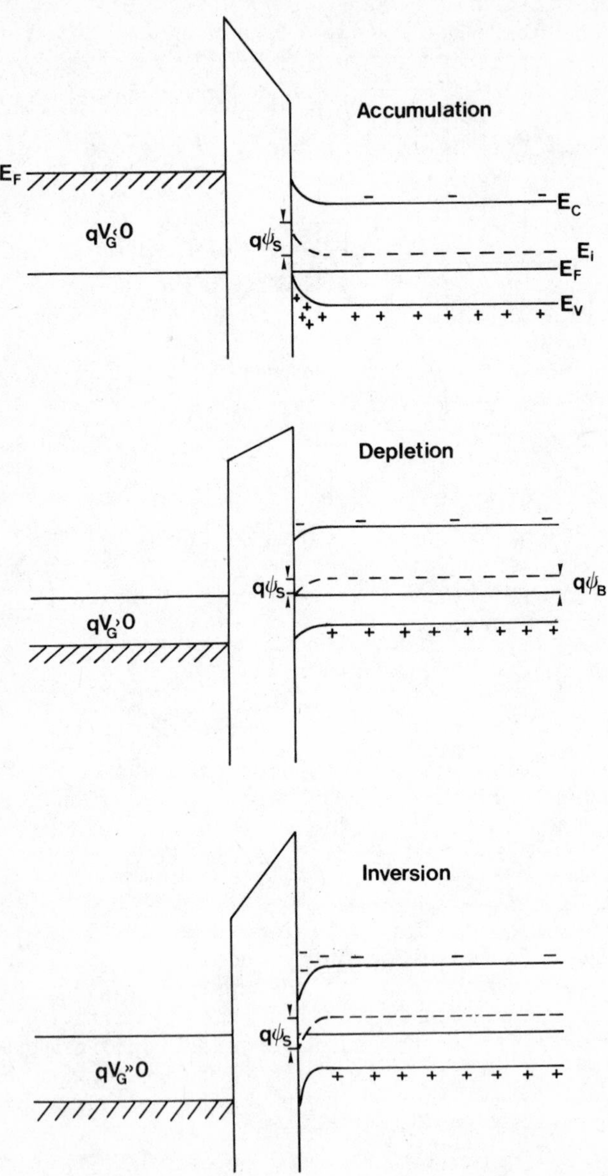

Figure 1.2 Energy band diagram near the surface of a p-type semiconductor for different applied bias conditions.

potential is simply solved in this ideal case by evoking the requirement for overall charge neutrality, that is, the charge on the plate must equal the sum of the charge in the inversion and depletion regions. If V_G is the applied voltage and V_{ox} is the potential across the insulator, then, in the absence of any work function difference

$$V_G = V_{ox} + \psi_s \tag{1.3}$$

where

$$V_{ox} = \frac{Q_s d}{\varepsilon_o \varepsilon_{ox}} = \frac{Q_s}{C_{ox}} \tag{1.4}$$

Q_s is the charge in the semiconductor, d is the oxide thickness, and C_{ox} is the oxide capacitance per unit area.

As the band-bending is increased from zero, the surface depletion region penetrates deeper into the semiconductor to balance the charge on the gate until the inversion region forms; at this point the surface potential is effectively "pinned", since large variations in the charge in the inversion region can be accommodated by only small changes in ψ_s. The maximum depletion depth is given by:

$$x_d \text{ (max)} = \left[\frac{2\varepsilon_s \varepsilon_o \psi_s \text{ (inv)}}{q\, N_A}\right]^{1/2} \tag{1.5}$$

The capacitance-voltage (C-V) characteristics of the MOS system are the basis of the most important of the investigative techniques. The overall capacitance is basically the series combination of the oxide capacitance C_{ox} (= $d/\varepsilon_o\varepsilon_{ox}$) which is not a function of the voltage, and the capacitance of the semiconductor space-charge region. Figure 1.3 shows the general form of the C-V characteristics for a p-type semiconductor. Under strong accumulation (negative gate voltages) the semiconductor behaves essentially as a metal and only the oxide capacitance is seen. At the flat-band condition, the semiconductor capacitance, C_D, becomes

$$C_D \text{ (flat-band)} = \frac{(2\varepsilon_s \varepsilon_o)^{\frac{1}{2}}}{L_D} \tag{1.6}$$

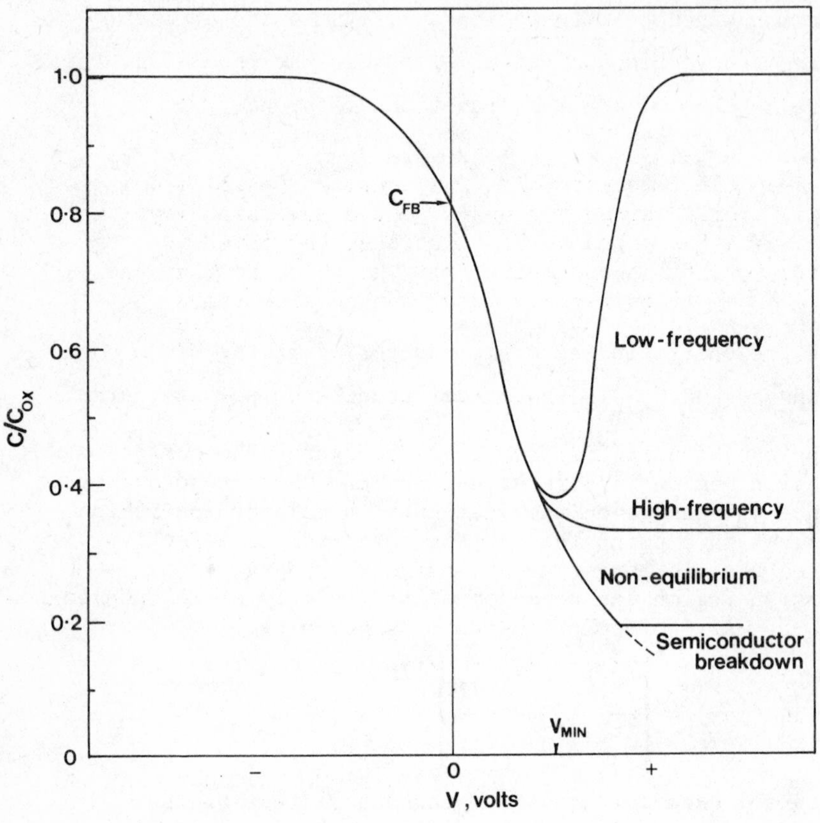

Figure 1.3 Capacitance-voltage curves for an idealized MOS structure (p-type semiconductor).

where $L_D = \left[\dfrac{2kT\varepsilon_o\varepsilon_s}{N_A q^2}\right]^{\frac{1}{2}}$ is the extrinsic Debye length for holes

so that the total flat-band capacitance is

$$C_{FB} = \frac{\varepsilon_o \varepsilon_{ox}}{d + \dfrac{\varepsilon_{ox}}{\varepsilon_s}\left(\dfrac{kT\varepsilon_s}{N_A q^2}\right)^{\frac{1}{2}}} \tag{1.7}$$

As the applied voltage increases and the surface depletion region deepens, the total capacitance falls as

$$\frac{C}{C_{ox}} = \frac{1}{\left[1 + \frac{2\varepsilon_o\varepsilon_{ox}^2}{qN_A\varepsilon_s d^2} \cdot V_G\right]^{\frac{1}{2}}} \qquad (1.8)$$

When the frequency of the measuring signal is low enough, mobile charge can drift to the inversion layer in response to the signal. In this case, the inversion region shields the depletion region from the gate, so that the capacitance climbs rapidly towards the oxide value, the capacitance minimum occurring at

$$V_G \text{ (strong inversion)} = \frac{Q_s}{C_{ox}} + 2\psi_B \qquad (1.9)$$

At higher frequencies, the inversion layer can no longer follow the gate signal, so that the capacitance continues to approach its minimum value of

$$C(\min) = \frac{\varepsilon_{ox}\varepsilon_o}{d + \left(\frac{\varepsilon_{ox}}{\varepsilon_s}\right) x_d(\max)} \qquad (1.10)$$

where $x_d(\max)$ is given by equation 1.5.

The transition between low and high frequency behaviour depends on the availability of minority carriers in the depletion region, and hence the low frequency curve can be extended to higher frequencies by the application of heat [12] or light, or by the proximity of appropriately biased diffusions (e.g. the source or drain of an MOS transistor).

If the dc bias is increased very rapidly, an inversion layer cannot form and the depletion region continues to extend into the semiconductor, until the breakdown field of the semiconductor is exceeded. The depth of the C_{min} dip is inversely dependent on the doping density of the substrate, as is shown by the curves of Goetzberger [11]. The normalized value of the flat-band capacitance (C_{FB}/C_{ox}) depends on the substrate doping and oxide thickness in a similar manner.

In a practical MOS system there are a number of factors which alter the ideal characteristics just discussed. Firstly, there is usually a work function difference between

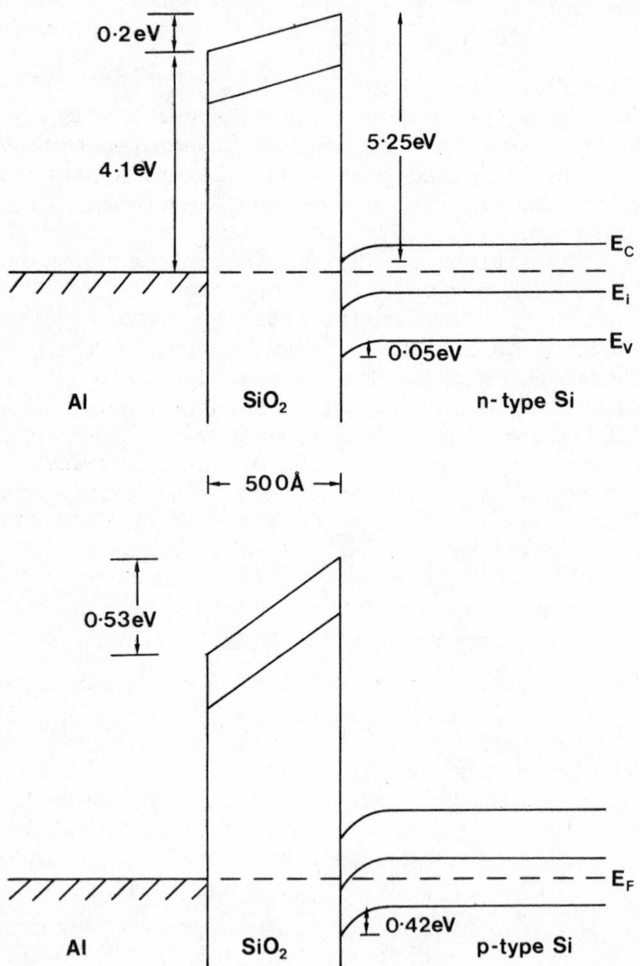

Figure 1.4 Band diagrams for Aluminium - Silicon dioxide - Silicon structures for 500 Å of SiO_2 and $N_D = N_A = 10^{16}$ cm^{-3}. No oxide charge assumed.

the semiconductor and the gate electrode. This results in a certain amount of band-bending with zero applied voltage, so that there is a parallel shift of the C-V characteristics along the voltage axis. The barriers for a variety of metals and semiconductor doping densities have been published by Deal and Snow [13]. The band diagrams for aluminium electrodes (by far the most widely used metal in commercial MOS integrated circuits) on 500 Å of SiO_2 over p- and n-type silicon ($N_D = N_A = 10^{16}$ cm^{-3}) are shown in Figure 1.4, assuming no surface states or oxide charge. As an alternative to aluminium, modern MOS integrated circuits sometimes use a deposited polysilicon layer, heavily doped to reduce its resistivity, as the top electrode. In this case the work function difference simply depends on the difference in positions of the Fermi level relative to the band edges in each region of semiconductor.

The second factor which effects the position of the C-V characteristics is fixed charge, either in the oxide or at the interface. Much confusion has been caused by some of the early workers in the field, who referred to this charge as "surface-state" charge because it is a property of the physical and chemical state of the interface. Great care should be taken to differentiate between this charge and that due to states at the interface which have energy levels in the semiconductor bandgap. The effect of fixed charge in the oxide depends on the position of the charge centroid, but the results of both charge types are usually described by an equivalent charge density of Q_f C/cm^2 located at the interface. For the Si-SiO_2 system Q_f is invariably positive, so that there is a parallel shift of the C-V characteristics in the negative direction of

$$\Delta V = \frac{Q_f}{C_{ox}} \qquad (1.11)$$

The origins of these charges will be discussed in later sections - they are to a large extent process-dependent.

A third difference between the ideal and observed characteristics of MOS devices is caused by the existence of surface states within the semiconductor band-gap, and localized at the semiconductor/oxide interface. These states are to a certain extent part of the basic physics of the interface, but they can be minimized by processing technology. Since the surface states are generally distributed in energy throughout the band-gap, and change their occupancy depending on the surface potential, they result in a smearing-out of the C-V characteristics, so that the magnitude of the threshold voltage required for surface inversion is increased for both p- and n-type silicon. Techniques for investigating surface states are described in Chapter 2 and the resultant device instabilities discussed in Chapter 3.

1.3 Field Effect Transistor Theory

Again, it is not intended to give a detailed description of the operation of MOS field effect transistors (MOSFETs) in this section, but a brief review is given as a reminder of the basic mode of operation of the device, and of its more important parameters. For fuller device descriptions, mainly of ideal, two-dimensional, devices, the reader is referred to the books by Sze [8], Grove [14], Crawford [15], or by Carr and Mize [16]. A diagramatic sketch of a basic n-channel MOST is shown in Figure 1.5.

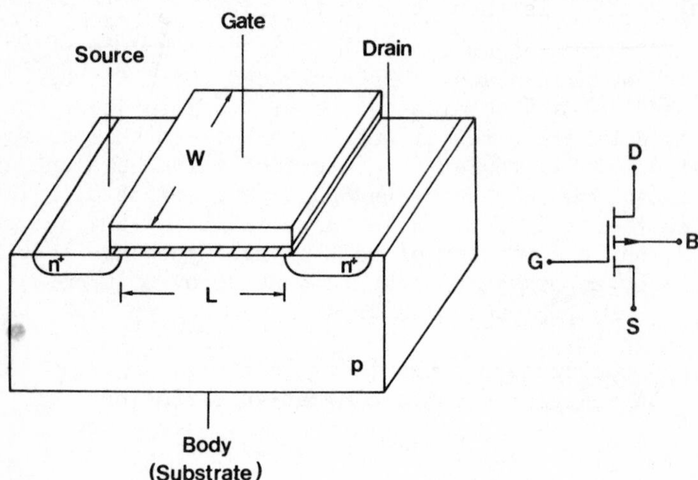

Figure 1.5 Schematic diagram and circuit symbol for an n-channel, enhancement type MOS transistor.

For gate voltages greater than the threshold voltage (Richman [17] gives theoretical values of threshold voltages) the inversion layer forms a conducting channel between the source and drain diffusions. For small drain voltages this channel acts as a voltage dependent resistor. In this unsaturated region (sometimes termed the "triode region") of the device characteristics the drain current I_D is given by

$$I_D = \beta [(V_G - V_T)V_D + \tfrac{1}{2}V_D^2] \tag{1.12}$$

where β is the gain factor, which depends only on the device geometry and the mobility of the carriers in the inversion layer;

$$\beta = \frac{W}{L} \frac{\varepsilon_o \varepsilon_{ox}}{C_{ox}} \mu_n, \tag{1.13}$$

and is usually expressed in $\mu A/V^2$. The drain current increases with drain voltage according to Equation 1.12 until the field in the oxide at the drain end of the channel is no longer sufficient to sustain inversion (the gate and drain voltages have the same sign). This occurs at

$$|V_D| \geqslant |V_G - V_T|. \tag{1.14}$$

The channel is then said to be "pinched-off" and the device operates in the saturation region with the drain current dependent, to a first approximation, only on the gate voltage, ie

$$I_D = -\frac{\beta'}{2} (V_G - V_T)^2 \tag{1.15}$$

The gain factor now has a value β′, which accounts for the shortening of the channel caused by the increasing reverse bias on the drain junction. These conditions prevail until the drain voltage reaches a value at which breakdown occurs, either by junction breakdown to the substrate, or by "punch-through" of the source and drain depletion regions.

The above description is, of course, very much simplified. In particular, it has not attempted to describe the drain current which flows for gate voltages below threshold (the subthreshold region, as described by Troutman [18]). This parasitic current is often a problem in integrated circuit design, since it can flow in nominally "off" transistors, and can result in leakage through unintentional parasitic or field-oxide transistors. Also, the above treatment has dealt with a two-dimensional model, and has neglected fringing fields - these factors are becoming increasingly important as modern transistors become smaller in the search for higher packing densities. Much effort is being made to accurately model the characteristics of small geometry devices [19].

1.4 Preparation Techniques

The most crucial stage in the production of any MOS device is the formation of the gate insulator. The requirements of this dielectric are very stringent; it must withstand high electric fields, have a high insulation resistance, its physical properties must match those of the semiconductor, it must have a low charge density, be impervious to charged species and to dopant atoms, and it must passify the semiconductor surface. Silicon dioxide films on silicon meet almost all of these requirements. The films may be produced in a number of ways; by thermal growth, anodization, sputtering, melting or sintering or by chemical vapour deposition (CVD) [20]. In practice, however, the films with a high degree of perfection required for gate insulators in MOS transistors are generally produced by thermal growth. Some of the other methods are also employed in other stages of integrated circuit manufacture, since they offer advantages of faster growth for thick ("field") oxides, they are lower temperature processes and, in the case of CVD oxides, they do not consume silicon from the substrate and so lead to flatter surfaces. Thermal oxides with thickness of the order of 1000Å are normally produced by heating the silicon slice in a quartz or silicon furnace (often double-walled) to a temperature in the range 950-1200°C with an ambient of either dry oxygen (for "dry" oxides) or steam ("wet" oxides) for periods in the range $\frac{1}{4}$ - 2 hours. Steam grown oxides grow considerably faster than dry ones but are not normally as stable, so that they are usually only employed for thick ("field") oxides. A third possibility for the gas ambient sometimes employed for the growth of gate oxide is oxygen bubbled through water. This technique has a growth rate between those involving dry and "steam" ambients; for all techniques the growth rate is a function of the substrate orientation, increasing for the (100), (110) and (111) faces respectively [21]. The electrical properties of the oxide are critically dependent on the furnace temperature, gas flow rates, rate of insertion and withdrawal from the furnace, and on the presence of any impurities, as well as on the conditions of the annealing treatment (usually heating in N_2 or H_2 in the temperature range 350-1000°C) normally performed after growth [22]. Thus the properties of an oxide produced by one manufacturer are still not readily reproducible by another manufacturer using nominally identical processing.

Although the properties of MOS devices are very heavily dependent on the oxide growth conditions, it must be remembered that the production of integrated circuits involves many other processing steps which will influence the properties of the finished device. For example, the oxide may be exposed to other high-temperature processes, and to dopant impurities, especially for self-aligned polysilicon-gate technologies. The oxide may also be exposed to radiation, either unintentionally during electron-beam evaporation of the metallization layer or during ion-implantation of dopants, or intentionally during x-ray or electron-beam lithography. The semiconductor manufacturer already has a large arsenal of MOS processing technologies, and new variations such as V-MOS (produced by anisotropic etching), DMOS (double-diffused to give short-channel transistors) and double-layer polysilicon techniques, generally tend to increase processing complexity in search of higher performance devices. Thus there remains a need to assess the characteristics and long-term relaibility of devices produced using these technologies since instability mechanisms originally overlooked will become increasingly important as transistors deviate further from the large, two-dimensional, "gradual-channel" devices discussed in the previous section.

1.5 Oxide Structure

The structure of thermally grown silicon dioxide has been extensively reviewed elsewhere [6, 21, 23]. It has an amorphous structure, although it can devitrify under certain conditions [24, 25]. Oxidation of silicon occurs without localized nucleation, so that very uniform thin films may be grown. The oxidation proceeds by the inward diffusion of oxygen to the interface, and since the molar volume of noncrystalline SiO_2 is about 2.2 times larger than that of silicon the grown oxide film is protective. Although it is almost perfectly amorphous, SiO_2 shows a high degree of short-range order [21], with each silicon atom being tetrahedrally surrounded by four oxygens; in contrast with crystalline forms the bond angles do not have a well defined value, but show a distribution about a mean value. This structure results in a very high degree of interface perfection and also gives a high dielectric strength ($\sim$ 9 MV/cm), but the rather open network leads to a high mobility for any ionic impurities.

The electrical properties of MOS devices are critically dependent on the structure of the Si/SiO_2 interface, but surface analysis techniques with sufficient sensitivity to investigate this structure have only recently become available, and there is still considerable discussion on the correct interpretation of the experimental findings. In the earliest papers, using auger analysis and argon ion sputtering [26] and ESCA [23], it was concluded that the interface region is 30-40Å thick, and is characterized by a natural interface roughness, with inclusions of silicon in the oxide matrix. With this model, in which there is no chemical form of Si other than SiO_2 and free Si itself, the electrically observed positive oxide charge was attributed to bonding defects between the two phases (Si and SiO_2) rather than to oxygen vacancies as had traditionally been assumed.

More recent experiments, using high-resolution electron microscopy [27, 28], auger analysis [29, 30, 31] and He backscattering [32, 33] have shown that the interface is much nearer to the ideal case than had previously been thought. Although TEM examinations of oxide defects by Irene [34] have detected silicon inclusions (which were worse of HCl-grown oxides), other authors [28] have been unable to find any protruberance larger than 4Å. Also, the thickness of the transition or interface region is now believed to be one or possibly two monolayers of SiO_x, where $0 < x < 2$; that is, less than 10Å. There may, however, also be some restructuring [33] of the top one or two layers of the silicon substrate. The number of disordered atoms in the interface is about a factor of two less for oxides grown on (100) surfaces compared to those grown on substrates with (111) orientation [32], which correlates well with the variations of fixed charge, but there is an inverse relationship between the interface width (which increases with increasing oxidation temperature [31]) and the fixed charge, which has not been explained. As well the localized thickness of the transition region, the interface is also characterized by long-range fluctuations in its position, so that an overall view may look like the schematic of Figure 1.6.

The experimental techniques used to examine the structure of the interface are not sensitive enough to detect the presence of any foreign atoms. It is known that hydrogen incorporation can have dramatic effects on the electrical properties of the oxide, but the form in which it is bonded at the interface or in the bulk is still unclear, although Revesz [35] has discussed observations (using internal infrared absorbtion spectroscopy) of both SiH and SiOH groups. Other authors [36, 37, 38] have produced theoretical models of the origins of the various oxide charges, but it is not certain whether they are caused by imperfections in the Si-O bonding and non-stoichiometry or by the presence of other atoms.

1.6 Instability Mechanisms

Deviations from the ideal MOS behaviour described in

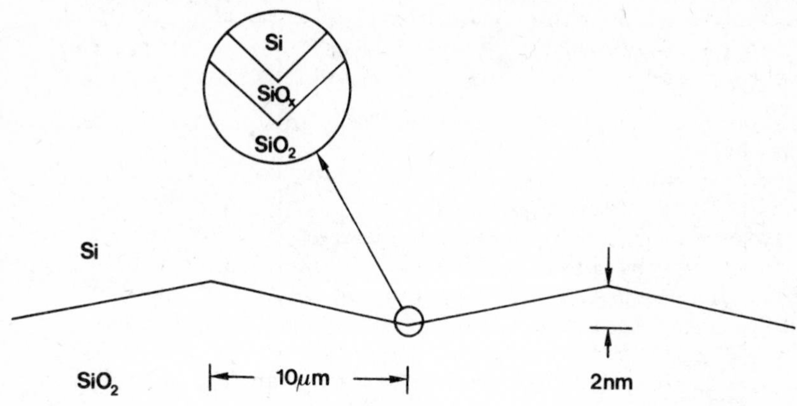

Figure 1.6 Schematic Si/SiO_2 interface, as suggested by auger studies (After Helms et al [31]).

Sections 1.2 and 1.3 are caused by charges in the oxide or at the semiconductor interface. In the early stages of MOS development confusion was created by the existence of a number of mechanisms capable of affecting the device characteristics. These mechanisms were not always referred to in the same terminology, so that it was not always obvious whether two workers were dealing with the same phenomenon or not. As a further confusing factor, the mechanisms are often interdependent, as well as being a function of the detailed preparation conditions employed. The most important classifications of the various oxide charge mechanisms are depicted in Figure 1.7. The terminology for the major types of charge, used in Figure 1.7 and in the remainder of this book, is that proposed by Deal and a committee of the Electrochemical Society and the IEEE (J. Electrochem. Soc, April 1980 and IEEE Trans. Electron Devices, March 1980).

Surface states are states within the forbidden gap of the semiconductor, localized at the Si/SiO_2 interface, and are partly due to the loss of lattice periodicity. For cleaved semiconductor surfaces the density of surface states is of the same order as the number of surface atoms, and it is that fact that prevented Shockley and Pearson from fabricating useful field-effect transistors. Happily, thermal oxidation of silicon reduces the surface state density

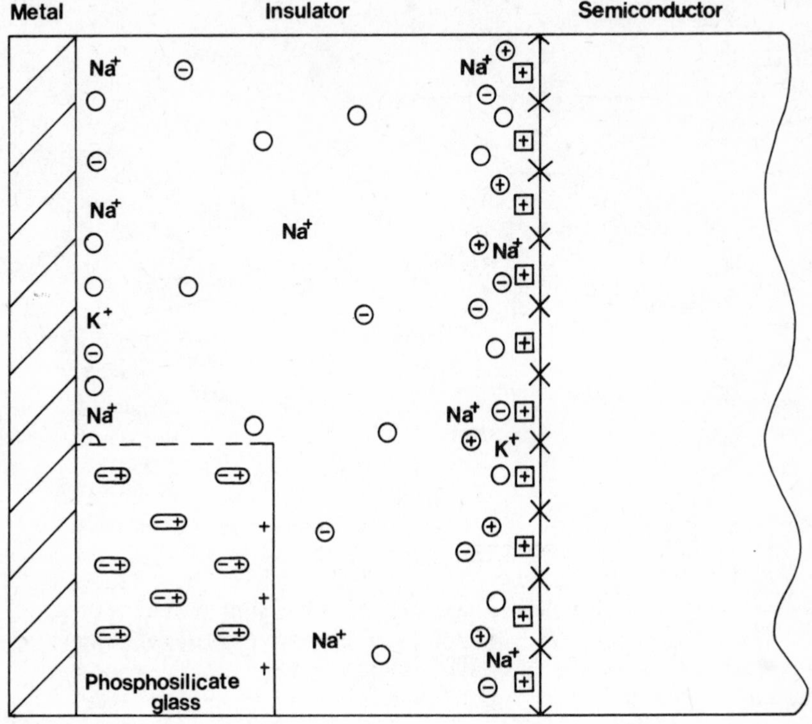

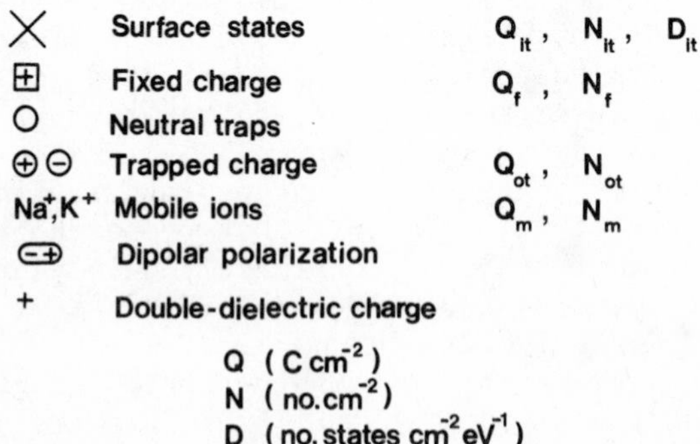

Figure 1.7 Basic classification of charge centres in a non-ideal MOS structure.

by several orders of magnitude, so that practical devices may be produced. Although surface states cause a deviation in device characteristics from the ideal case, they do not in themselves give rise to instabilities, rather it is the growth of surface state densities during life which can cause unreliability. Surface states can be either donors or acceptors, and can exchange charge with the semiconductor conduction and valence bands with a characteristic time constant, so that historically they have been characterised as "fast states" and "slow states". For practical MOS devices, the states resulting from energy levels localized at the Si/SiO_2 interface (and which really should be termed "interface" states, to distinguish them from those at the SiO_2/air interface) are regarded as "fast states", although paradoxically the time constant can be long, especially at low temperatures. The presence of surface states results in an increase (in magnitude) of the dc MOS transistor threshold voltage compared to the theoretical value, although for sufficiently fast ac measurements the two values coincide. Surface states are discussed in Chapter 3.

The relatively open structure of amorphous silicon dioxide results in a high mobility for many impurities. These impurities, if charged, will alter the electrical characteristics of the system, and will drift during bias-temperature ageing, or, more slowly, in service conditions. The mobile ions found in the Si/SiO_2 system are usually due to contamination by (positive) metallic ions, although ions from organic origins are sometimes observed. The causes and characteristics of mobile ions are considered in Chapter 4.

Another source of positive charge in the oxide is classified in Figure 1.7 as "fixed charge". Here, "fixed" refers to the position of the charge, which is usually within ~80Å of the interface, and also to the fact that the charge is independent of the surface potential (in contrast to the surface state case). However, the magnitude of this component of oxide charge is varied by different annealing conditions, and possibly by ageing, as it is a function of the structure of the interface.

The orientation of permanent dipoles, which may be present in the oxide or, more likely, in other gate insulators, in line with an applied field is shown in Figure 1.7 as dipolar polarization, and the resultant instability is discussed in Chapter 5.

Silicon dioxide is a wide bandgap insulator but several trap levels exist within the forbidden band which may become ionized during life, causing instabilities in device characteristics. These traps may occur throughout the thickness of the oxide, although it is usually only those near the interfaces which become ionized, except during exposure to radiation. The traps may become either negatively (electron trapping) or positively charged (the ubiquitous "slow hole trapping"). Charge trapping is considered in Chapters 6 and 7.

To many investigators, "instabilities in MOS devices" is synonomous with "threshold voltage drift". Certainly, changes in MOS transistor threshold voltage during life are the most obvious and important reliability hazard, but, more generally, any change in the electrical properties of the MOS system may be regarded as an instability. Thus the transistor gain (β) is affected by the carrier mobility in the channel, and its subthreshold characteristics are dependent on surface state densities. Two other instabilities which are important to long-term device reliability are related to the I-V characteristics of the insulator. Firstly, changes in leakage conduction through the insulation could affect the operation of dynamic digital integrated circuits or high-impedance analogue circuits and will degrade the storage in MNOS and FAMOS Read Only Memories. Secondly, changes in oxide structure or charge can result in dielectric breakdown of the insulator, which is required to withstand fields which are an appreciable fraction of the intrinsic dielectric strength during normal service conditions. This final instability is unusual in that it results in catastrophic failure of the device or circuit. These "instabilities" are considered in Section 8.

Although this book is primarily concerned with instabilities in MOS transistors and integrated circuits, it should be remembered that oxidised silicon surfaces are found in many other electronic devices, so that their reliability may be similarly affected.

2

Investigative Techniques

Numerous experimental techniques have been employed to investigate and characterize the MOS system. The most widely used test vehicle is the MOS capacitor, although the use of a transistor does sometimes increase the range of measurements which can be made. The techniques broadly fall into two categories; those which give an overall picture of the total charge stored in the oxide (such as the basic C-V method, and measurements of transistor parameters) and those which attempt to isolate individual charge storage mechanisms. This chapter is concerned only with techniques involving the measurement of the electrical parameters of an MOS device, and not with the techniques used to study its physical structure.

2.1 C-V Methods

2.1.1 Basic Method

The derivation of the ideal capacitance-voltage (C-V) characteristics of the MOS system was described in Section 1.2. Comparisons of experimental C-V curves with theoretical ones, either directly or by means of the more complicated analyses described in the forthcoming sections, are by far the most widely used techniques for investigating charges in MOS devices.

For oxide charge which does not alter with the surface potential of the semiconductor, the net magnitude of the charge can be calculated directly from the amount of the horizontal (along the voltage axis) shift between the theoretical and experimental curves. For cases where other charges which do change with surface potential are present, the C-V curve does not show a completely parallel shift, and so the voltage difference at the flatband capacitance (ΔV_{FB})

is used to calculate the "fixed" charge density. Although this is an extremely simple method it suffers from the fact that it results in only a single equivalent charge density, and cannot distinguish between any of the possible causes described in Section 1.6. However, when monitoring an MOS integrated circuit production line, this equivalent charge gives important information on the overall stability of the process. The charge centres may be distributed throughout the oxide, but, since this method cannot give information on the charge centroid, the equivalent fixed charge (Q_f) is taken as that charge density which, located at the oxide-semiconductor interface, would result in the same shift in flatband voltage:

$$Q_f = C_{ox} \Delta V_{FB} \tag{2.1}$$

2.1.2 Differentiation Method

The presence of surface states in an MOS system results in a change of shape of the C-V curve, so that analytic techniques slightly more complicated than simple inspection must be employed. Information is required about the total number of surface states, their distribution within the forbidden gap, and their time constants. The earliest method of extracting this information from C-V data is due to Terman [39], and involves a graphical differentiation as the technique produces the integral of the surface state density. The C-V characteristics of the sample are first plotted at a frequency which is high enough not to allow the surface state occupancy to follow the measuring signal: this results in the "ideal" high frequency curve shown below in Figure 2.1.

The characteristics are then replotted at a lower frequency, and it is found that there is a "smear-out" due to the surface states. The total charge in the surface states (Q_{it}) at a given applied voltage is then given by

$$Q_{it} = C_{ox} \Delta V \tag{2.2}$$

where ΔV is the voltage shift. The number of surface states per unit energy (D_{it}) is then

$$D_{it} = \frac{1}{q} \left(\frac{\partial Q}{\partial \psi_s} \right)_V \quad \text{states/unit area/ eV} \quad \ldots\ldots \tag{2.3}$$

One of the major disadvantages of this method is the uncertainty in the relationship between the surface potential ψ_s and the applied voltage. Also, the shift between the two curves only becomes easily measurable at large surface state

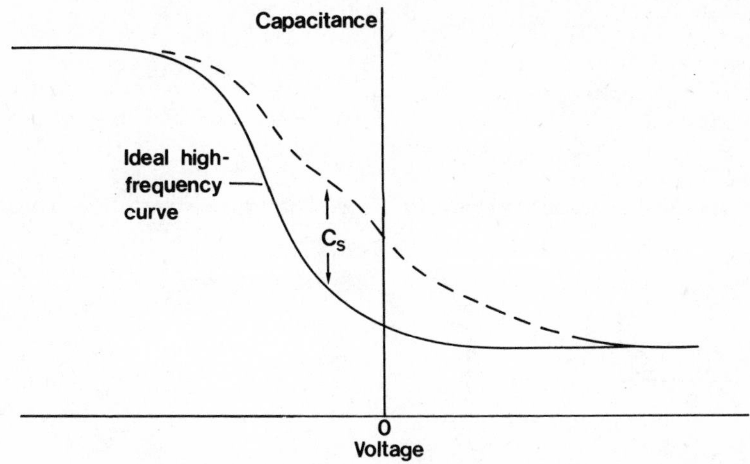

Figure 2.1 High and low frequency C-V plots for an MOS capacitor on p-type silicon, showing "smear-out" due to surface states.

densities. It is, in theory, possible to measure the time constants of the surface states by repeating the low frequency curve at a number of frequencies.

2.1.3 Integration or "Quasistatic" Method

The disadvantages of the above differentiation method can be partially overcome by an "integration" method due to Berglund [40]. This method uses the low-frequency "quasi-static" C-V plot, in which both the measuring frequency and the ramp rate are slow enough for the surface states to be always in equilibrium. Under these conditions it can be shown that

$$\frac{\partial \psi_s}{\partial V} = 1 - \frac{C}{C_{ox}} \qquad (2.4)$$

On integration, this gives

$$\psi_s(V_1) - \psi_s(V_2) = \int_{V_2}^{V_1} \left[1 - \frac{C}{C_{ox}} \right] . \, dV \qquad (2.5)$$

which allows the surface potential to be calculated within an additive constant at any applied voltage. This constant is evaluated by comparing the actual and theoretical curves in the accumulation region. We also have the relationship

$$\frac{d\psi_s}{dV_{ox}} = \frac{C_{ox}}{C} - 1 \tag{2.6}$$

which, together with Equation 2.5 allows a curve of $\partial\psi_s/\partial V$ versus ψ_s to be drawn directly from the quasistatic C-V plot. Now

$$\frac{\varepsilon_o \varepsilon_{ox} V_{ox}}{d} = q \int_0^\infty D_{it} F(E_T).dE + Q_s \tag{2.7}$$

where $F(E_T)$ is the value of the Fermi function at the energy (E_T) of the surface states and Q_s is the charge in the semiconductor. Differentiation of Equation 2.7 gives

$$\frac{\partial\psi_s}{\partial V_{ox}} = \frac{\varepsilon_{ox}\varepsilon_o}{d} \Bigg/ \left[\frac{dQ_s}{d\psi_s} + q\, D_{it}\, (q\psi_s)\right] \tag{2.8}$$

Since $dQ_s/d\psi_s$ may be calculated if the doping density of the semiconductor in the surface region is known, this equation may be compared with the results of the quasistatic C-V plot to give the surface state density as a function of ψ_s.

One practical problem with the quasistatic method is the actual measurement of the C-V characteristics at very low frequencies. Even with a phase-locked amplifier, measurements with sinusoidal signals are difficult below 5 Hz. Instead, Kuhn [41] has used the method of measuring the displacement current flowing through the MOS capacitor when a linear voltage ramp is applied, using the circuit shown in Figure 2.2. Using a high impedance amplifier, sweep speeds as low as 5 mV/sec may be used, limited by the oxide conductance and minority carrier lifetime in the substrate. Under these conditions the resulting C-V plots are independent

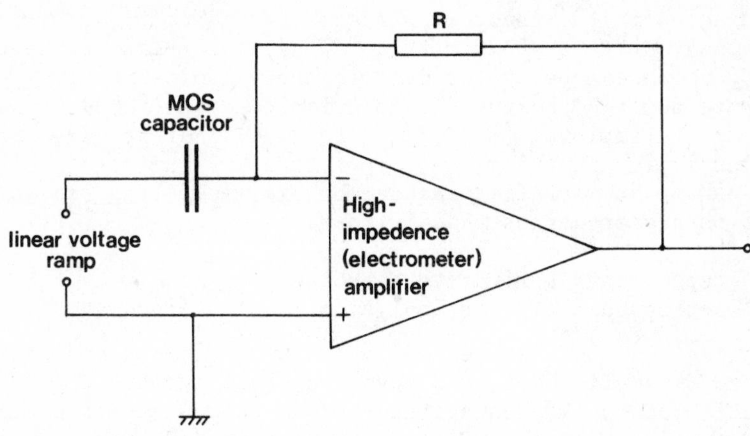

Figure 2.2 Basic circuit required for quasi-static C-V measurements (After Kuhn [41]).

of sweep speed or polarity, and the detection sensitivity is of the order of 10^{10} states/cm^2/eV near mid-gap.

The measurement of the surface state distribution in metal-nitride-oxide-silicon (MNOS) devices is complicated by the buildup of charge at the interface between the two insulators. A method of overcoming this problem is due to Arnold and Schauer [41a], and involves superimposing a fast voltage-ramp on a slow one.

2.1.4 Temperature Method

A method of extracting surface state information from high frequency C-V plots is due to Gray and Brown [42, 43]. In their method the position of the Fermi level, and hence the occupancy of the surface states, is altered by changing the temperature of the sample. The flat-band voltage is recorded as a function of the temperature by constantly monitoring the capacitance. Since, by definition, the bands are flat up to the surface in this condition, the surface potential is equal to the bulk potential, the latter being calculable from the temperature and the doping density. The surface state density can thus be calculated from the differential flat-band voltage. Oxide space charge and electron affinity differences do not affect the measurement as long as they are temperature independent, but it is necessary for the measuring frequency to be high enough for the surface states not to follow it. Actually, it has recently been shown [44] that variations in the semiconductor work function with temperature can introduce large inaccuracies in the measured surface state densities. The main disadvantage of this method is that the region of the bandgap which can be explored is severely limited by the substrate doping density (and polarity) and the temperatures which may be employed (usually 77 to 300K). Great care must be taken in interpreting low temperature C-V curves since the minority carrier lifetime can be extremely long and an inversion region cannot readily form.

2.1.5 Capacitance Derivative Method

A method suitable for analysing small surface state densities (10^{10}-10^{11} states/cm^2/eV) around mid-gap has been reported by Amelio [45], and involves measuring the derivative of the MOS capacitance ($C' = dC/dV$) as a function of the applied voltage. The measurement is made by monitoring the second harmonic of an ac signal as the bias voltage is slowly ramped from accumulation to inversion. The frequency of ac signal used is high compared with the minority carrier lifetime but low enough for the surface states under investigation to respond.

Qualitatively, the capacitance derivative data has the same appearance as that obtained by the conductance technique (described in Section 2.2), showing a pronounced peak around the flatband voltage. Figure 2.3 depicts typical C′-V data for a particular oxide thickness and substrate doping density. Surface state densities which are constant across the band-gap result in a reduction in peak height, as is indicated in the diagram; for curves of this type the surface state density is tabulated [45] as a function of the reduction in peak height. When the surface state distribution has a finite structure in the band-gap the C′-V plot exhibits a secondary peak at the appropriate surface potential. To extract quantitative information from these plots is also possible but requires considerably more effort as the curve must be regressively fitted to a theoretical expression. As well as an inability to resolve fine detail in the surface state distribution this method suffers, in common with most other methods, from inaccuracy when random spatial fluctuations in oxide charge exist. The fluctuations result in an overall smoothing of the C′-V peak, and although in theory an allowance may be made for this effect it involves a largely intuitive choice of a constant to represent the characteristic area with no fluctuations of oxide charge.

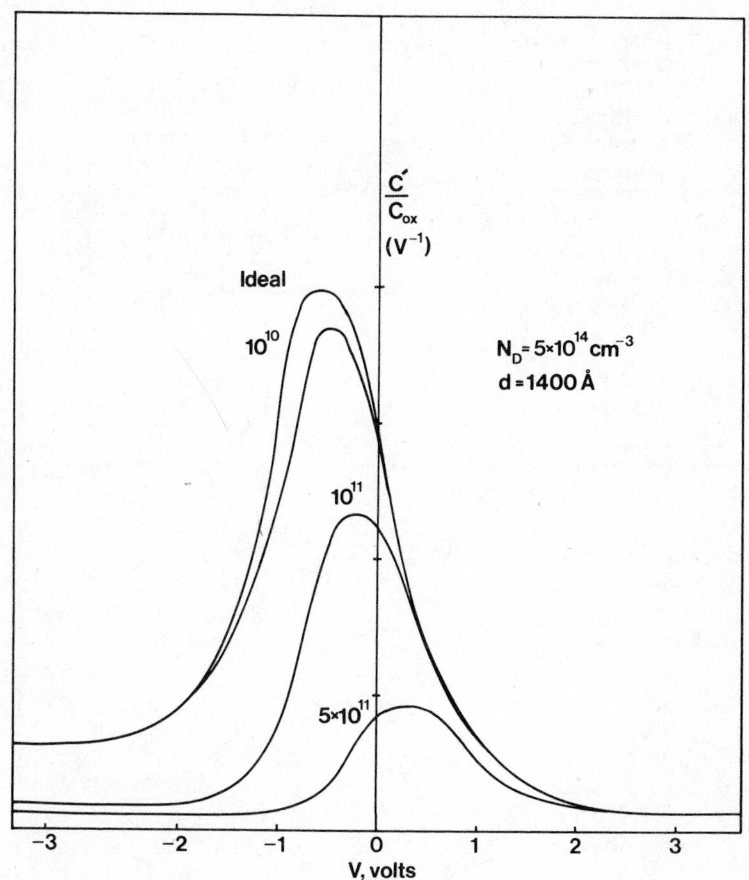

Figure 2.3 Typical capacitance derivative spectra (C′-V) for various surface state densities. (After Amelio [45])

2.1.6 Transient Capacitance Method

Wang and Evwaraye [46] have reported a technique for measuring surface states in MOS transistors by pulsing the surface from deep depletion to accumulation, and measuring the transient capacitance when the surface is restored to deep depletion. This method, sometimes known as deep-level transient spectroscopy (DLTS), is also capable of detecting bulk states in the semiconductor. Referring to the C-V curves of Figure 2.4, the solid line shows the normal deep depletion

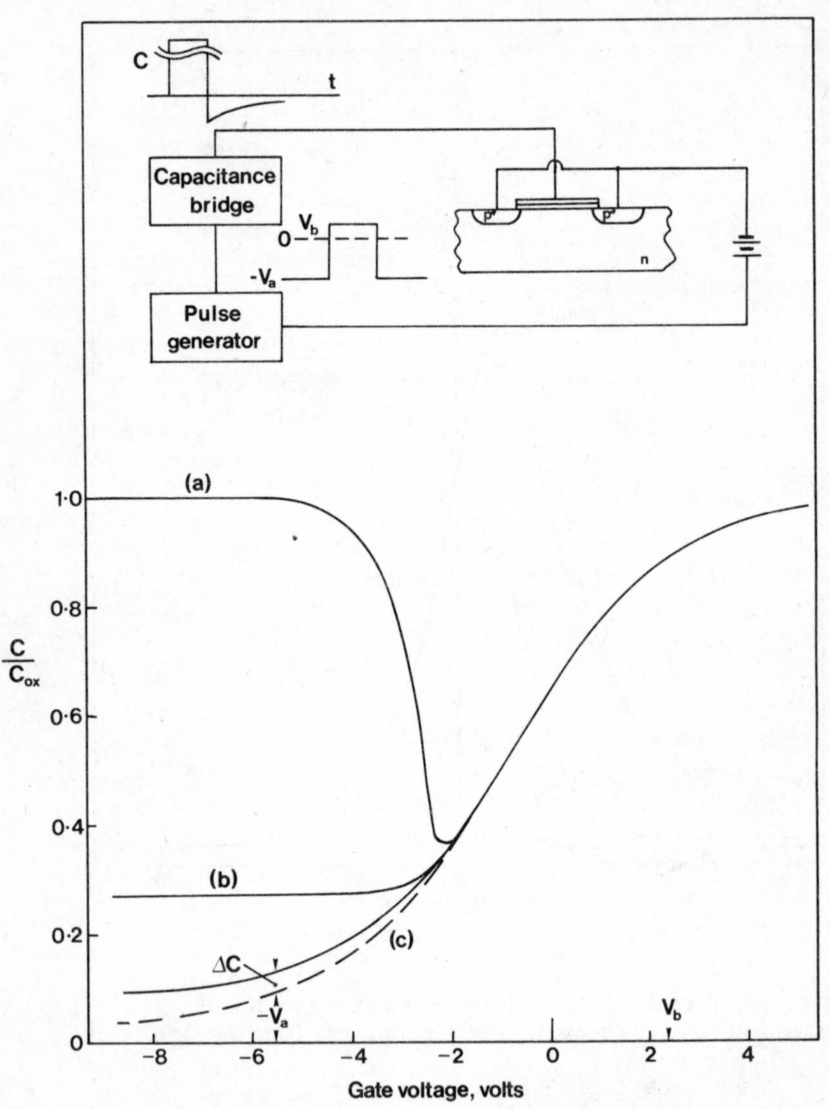

Figure 2.4 Curve (a) - equilibrium low frequency C-V plot; curve (b) - high frequency C-V plot in inversion; curve (c) - deeply depleted case. Inset: schematic circuit to measure the emission of electrons from acceptor surface states. (After Wang and Evwaraye [46]).

case. When the gate is pulsed to V_b the surface is biased into accumulation and electrons are trapped in the surface states (assumed here to be acceptors). If the gate voltage now returns to $-V_a$ the trapped electrons will cause the depletion capacitance to be greater, and hence the capacitance is C_a' as shown by the dotted line. Electrons will be emitted from the surface states with an emission rate which depends exponentially on temperature, so that the capacitance will decay back to the normal case. For a single-level surface state the decay will also be exponential, so that the time constant may be obtained by measuring the value of ΔC between two times after the return to deep depletion. Wang and Evwaraye achieved this by means of a two-channel boxcar detector with windows in the 0.1 to 50 msec region; the measuring signal was in excess of 25 MHz. By recording the value of ΔC with a given pair of windows as the temperature is varied from 100 to 325K a thermal scan similar to that shown in Figure 2.5 is obtained. In this case the sample has been implanted with gold, leading to the discrete energy levels shown by the structure of the thermal scan, and labelled A1-5. Thermal scans with different window times allow Arrhenius plots of emission rate from the surface state against reciprocal temperature to be made for each of the peaks, from which the energy level of each peak may be obtained. Capture cross-sections may also be calculated.

In order to measure surface states with energies in the lower half of the bandgap, minority carrier injection is accomplished by a short pulse which forward biases the source - and drain-substrate junctions whilst holding the gate voltage negative. This results in hole capture, so that ΔC is now negative. Unfortunately, this procedure can also result in the filling of electron traps with levels between deep depletion and inversion, so that care must be taken to correctly interpret the resulting false peaks in the thermal scan.

The transient capacitance method requires more sophisticated instrumentation than most of the other methods available for investigating surface states, but it does enable most of the band gap to be explored with a single sample, and it gives both the energy distribution and capture cross-sections of the states. Above all, however, its extreme sensitivity (10^8 states/cm^2/eV claimed by Wang and Evwaraye) will prove useful in investigating modern interfaces, and the changes introduced in them by later processing steps.

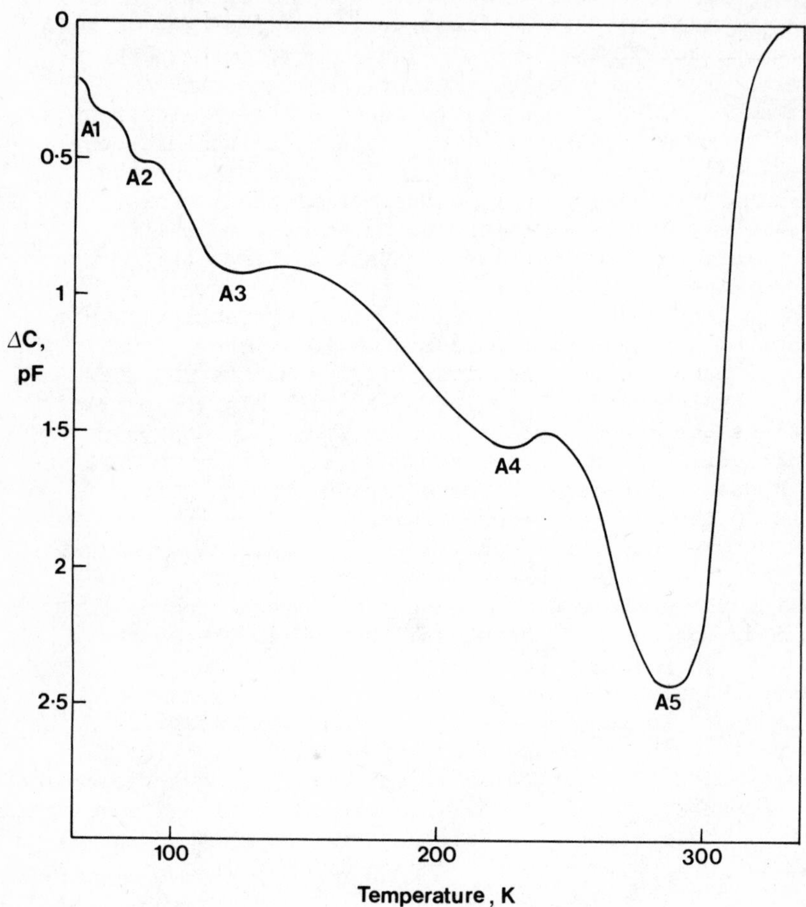

Figure 2.5 Typical transient capacitance thermal spectrum, using a two-channel boxcar averager with windows set at 0.15 and 0.5 msec after return to deep depletion: Signal frequency = 27.5 MHz. (After Wang and Evwaraye [46].)

2.2 Conductance Method

The problem with most of the C-V methods for investigating surface states is that the capacitance contribution due to the surface states is buried by the much larger oxide and depletion capacitances. This problem is overcome by the conductance method which was described in detail by Nichollian and Goetzberger (1967) in their classic paper [10], and which has a high sensitivity (10^9 cm^{-2}/eV), especially in the mid-gap region.

The theory of the conductance method is based on the assumed equivalent circuit shown in Figure 26(a) where ω is the angular frequency,

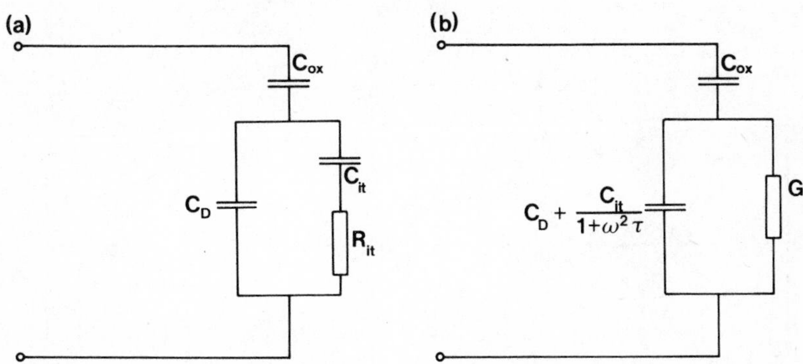

Figure 2.6 Equivalent circuits of MOS system with surface states. (After Nicollian and Goetzberger [10])

C_{ox} is the oxide capacitance, C_D is the semiconductor depletion capacitance and C_{it} and R_{it} represent the surface states with a time constant $\tau = C_{it}R_{it}$. This circuit can be converted to that shown in Figure 2.6(b), where the parallel conductance is due soley to the surface states. Measurement of the conductance may be performed with a normal bridge circuit, although care must be taken to limit the excursion of the surface potential due to the ac signal to a few kT/q, otherwise harmonics introduced by the voltage dependence of the capacitance will contribute a spurious conductance. Since the loss angle is small for low surface state densities another possible source of error is surface leakage; this may be minimised by keeping the sample in a dry ambient and, where it is possible to select the sample processing, by using a thin oxide so that the impedance due to C_{ox}, which appears in series with the surface state conductance, is minimized. Examples of conductance-voltage characteristics at two frequencies are compared in Figure 2.7 with the corresponding C-V curves; the increased sensitivity to surface states is clearly shown.

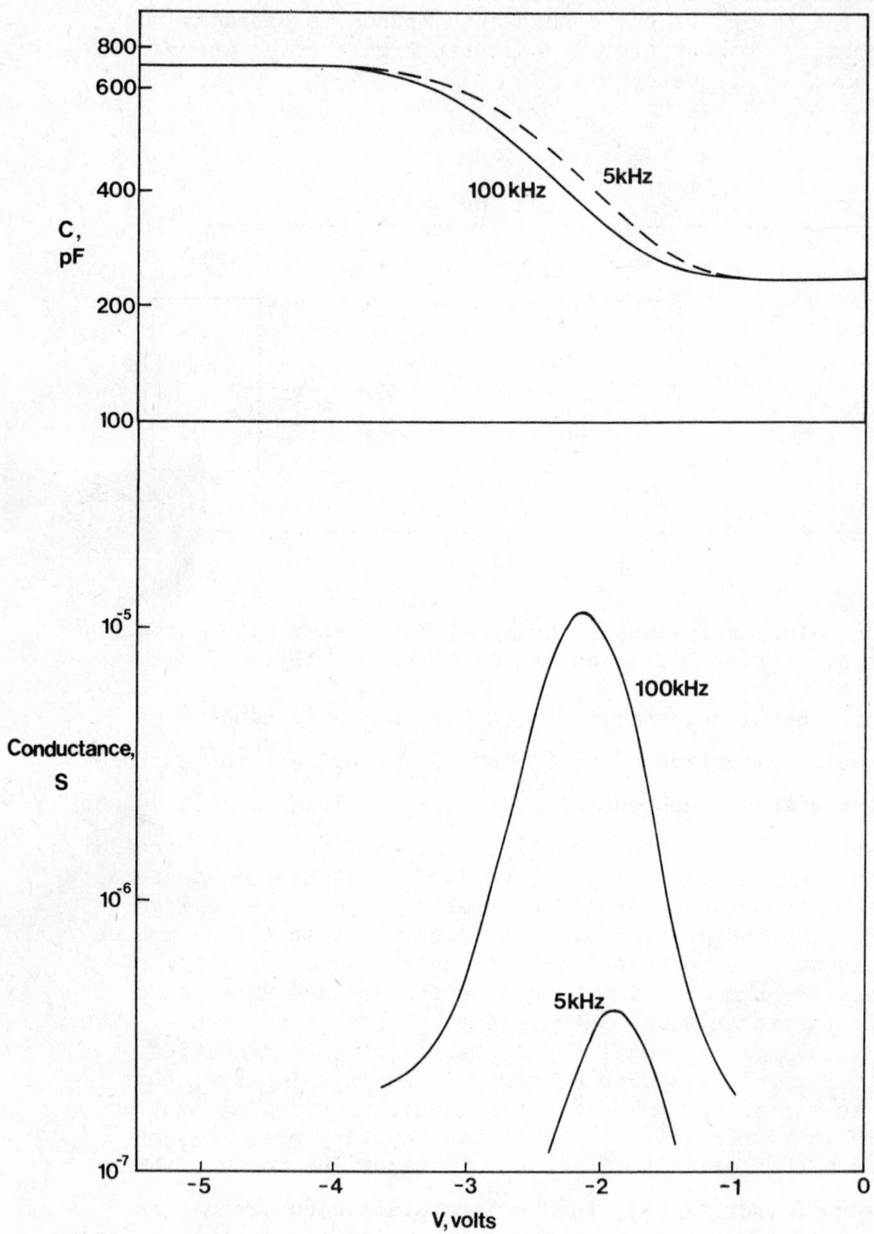

Figure 2.7 Comparison of MOS capacitance and conductance-voltage curves at two frequencies. (After Nicollian and Goetzberger [10])

Analysis of the conductance data proceeds by removing the reactance due to the oxide capacitance (measured under strong accumulation) from the equivalent parallel conductance of the surface state branch of the equivalent circuit. The resulting admittance, G_p, is given by

$$\frac{G_p}{\omega} = \frac{C_{it}\ \omega\ \tau}{1 + \omega^2\tau^2} \qquad (2.9)$$

This function peaks at $\omega\tau = 1$, so the time constant of the surface states at the surface potential corresponding to the value of the applied bias at the conductance peak may be calculated directly. Measurements at other frequencies yield the time constants at other energies in the band-gap. The value of this function at the peak is $C_{it}/2$, so that the surface state density may also be calculated.

The above description of the conductance is not universally valid, but is sufficiently accurate in the weak inversion region ($\psi_F < \psi_s < 2\psi_F$). There are two principal causes of deviations from the simple case. Firstly, the surface potential shows small localized fluctuations about its mean value due to variations in the oxide charge density*. Since it has been shown experimentally that the time constant τ is exponentially dependent on the surface potential, small variations in ψ_s lead to a large time-constant dispersion. These fluctuations are most significant in the accumulation and depletion regions. The second reason for deviations from the simple theory is that the surface states exist as a continuum across the bandgap, and not at a single energy. This again results in a time-constant dispersion. By considering the surface state continuum and also assuming a Poisson distribution of oxide charge Nicollian and Goetzberger [10] have derived a statistical expression for the conductance curve which may be fitted to experimental curves obtained in the accumulation and depletion regions. Unfortunately, however, this expression contains a factor A_c which is the small area over which the surface potential is considered to be constant; the exact value of this constant must be obtained by curve fitting, although its value is roughly proportional to the square of the depletion depth at ψ_s.

In the mid-gap region the equivalent circuit must again be modified; each value of C_{it} will have two resistances

*Large fluctuations in surface potential have recently been discovered [47, 48], and will be considered in Chapter 3.

associated with it, corresponding to the capture resistances for holes and electrons. Finally, in the strong inversion region the effect of the surface states is effectively short-circuited by the resistances associated with the transport of carriers through the bulk and the depletion region, since as the surface potential is "pinned" C_{it} is much smaller than C_D.

Thus although the conductance method is the most widely used technique for surface state measurements, analysis of the data is based on the use of an assumed equivalent circuit and involves extensive computation.

2.3 Charge Pumping

In order to investigate the stability of commercial MOS processes a technique for measuring surface states is required which can utilise small geometry transistors, with gate capacitances of ~1 pF, as the test vehicle. Neither the C-V nor the conductance method is suited to this situation, where stray capacitance from the encapsulation is of the same magnitude as the device capacitance, and the fine details of the manufacturing process are unknown. The phenomenon of "charge pumping" was originally described in 1969 by Brugler and Jespers [49], and developed into a simple and sensitive measuring technique for surface states by Elliot [50] and Owczarek and Kolodziejski [51].

The basic circuit used for charge pumping measurements is shown in Figure 2.8 for the case of a p-channel transistor (negative threshold voltage). When the transistor is switched "on" by a negative gate voltage holes flow from the source and drain regions into the channel. Some of these holes are trapped by surface states, and the rest remain as mobile charge in the inversion region. If the gate pulse is now removed this mobile charge drifts back to the source and drain regions, under the influence of V_R. The charge in the surface states remains however, to be neutralized by electrons from the substrate, and thereby gives rise to a unidirectional "pumped" current of majority carriers from the source and drain to the substrate, in a direction opposite to the normal junction leakage. For a repetitive gate signal, the charge pumping current (I_{CP}) can be written as

$$I_{CP} = fAq\Delta N_{it} + fA\alpha C_{ox}(V_T - V_G) \qquad (2.10)$$

where f is the pulse frequency, A is the gate area, ΔN_{it} is the number of surface states/unit area with energy levels which sweep through the Fermi level when the pulse is applied, and α is a fractional constant. The first term on the right

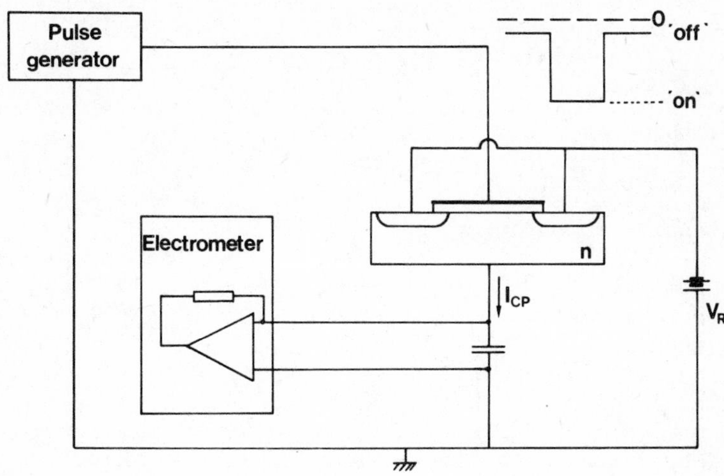

Figure 2.8 Basic charge pumping circuit. (After Elliot [50])

hand side of Equation 2.10 is the pumped current due to surface states, and the second is a small current which is termed the "geometric component" by Brugler and Jespers. This geometric component arises from a small fraction (α) of the mobile charge in the inversion layer which is unable to drift back to the source and drain diffusions, but recombines in the substrate instead. It is usually small for short transistors and pulse fall times greater than ~10 ns [51], and can be minimized by adding a reverse bias to the source and drain diffusions; this bias also causes a slight reduction in the saturated charge pumped current because of channel shortening. A typical set of charge pumping curves is shown in Figure 2.9 where it can be seen that, neglecting the geometric component of the $V_R = 0$ curve, the charge pumped current increases rapidly from zero when the pulse amplitude is sufficient to cause inversion, and saturates when the surface potential becomes pinned at $2\psi_F$. Thus a single measurement of the charge pumped current with a gate pulse taking the surface from inversion to accumulation, and with V_R large enough just to suppress the geometric component without causing excessive channel shortening (~1V), is sufficient to characterize the surface state density operative in the normal transistor operating region.

The charge pumping method is also capable of measuring the surface state density as a function of position in the band-gap, although once again the quasistatic C-V plot must be

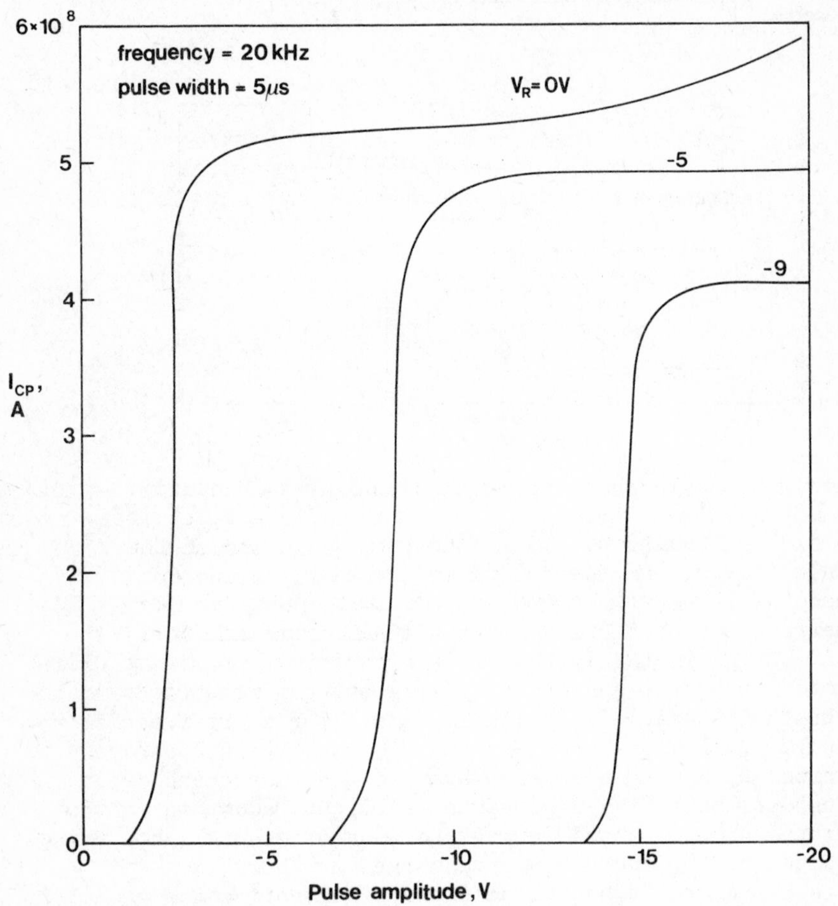

Figure 2.9 Charge pumping current as a function of gate pulse amplitude with source and drain to substrate reverse bias voltage (V_R) as a parameter. (After Elliot [50]).

employed to relate the applied gate voltage to the surface potential. In the method used by Elliot [50], the pulse amplitude is held constant and the charge pumping current recorded as a function of the pulse base level. This results in characteristics such as those shown in Figure 2.10; included in the figure are schematic band diagrams corresponding to selected points on the curve. The transistor has a threshold voltage of -2.5V, so that at A the surface potential is pinned at $2\psi_F$ and no surface states change their occupancy; the small I_{CP} seen is a geometric component reflecting the difference in the number of mobile charges in the inversion region between the pulse "on" and "off" levels. As the curve progresses through B and C the surface states are still filled during the pulse "on" period, but a progressively larger number of them are able to discharge in the pulse "off" period. Finally, at D the pulse "on" amplitude is insufficient to invert the surface, so that the charge pumped current falls to zero again. Although donor states have been used for this explanation the argument is not significantly altered if acceptor states are also present. It can be seen that the shape of the rising edge of Figure 2.10 contains information on the density of surface states as a function of surface state potential.

Charge pumping is an extremely valuable method of investigating surface states in the central portion of the bandgap, offering high sensitivity and the ability to handle small test transistors. It requires a minimal amount of equipment, but does rely on the quasistatic method if the distribution of surface states with energy is required. Information on the capture cross sections is not obtained. When a single value of I_{CP} is taken to characterize the growth of the total number of surface states between accumulation and inversion during bias-temperature stressing, it must be remembered that, if the transistor exhibits significant threshold voltage shift, fixed pulse levels will not result in the same region of the bandgap being swept by the surface potential.

2.4 MOST Characteristic Methods

Since one of the main reasons for investigating the Si/SiO_2 interface is to ascertain the effects instabilities will have on MOS transistor performance, it is clearly sensible to make direct measurements of the MOST characteristics. The disadvantage of this method is that it is often not possible to distinguish between instability mechanisms, so that the steps required to remove the instability from the manufacturing process are not usually discovered. The chief transistor characteristics which are monitored at intervals during a stressing sequence are the threshold voltage V_T, and gain factor, β, although the subthreshold and unsaturated region characteristics are usually also measured. Both V_T and β are normally obtained from the $V_G - \sqrt{I_D}$ curve in the saturated region. A number of precautions must be employed during the measurements if the results are to be reproducible. For example, measurements should be made starting at the high current end of the characteristics, so that self-heating

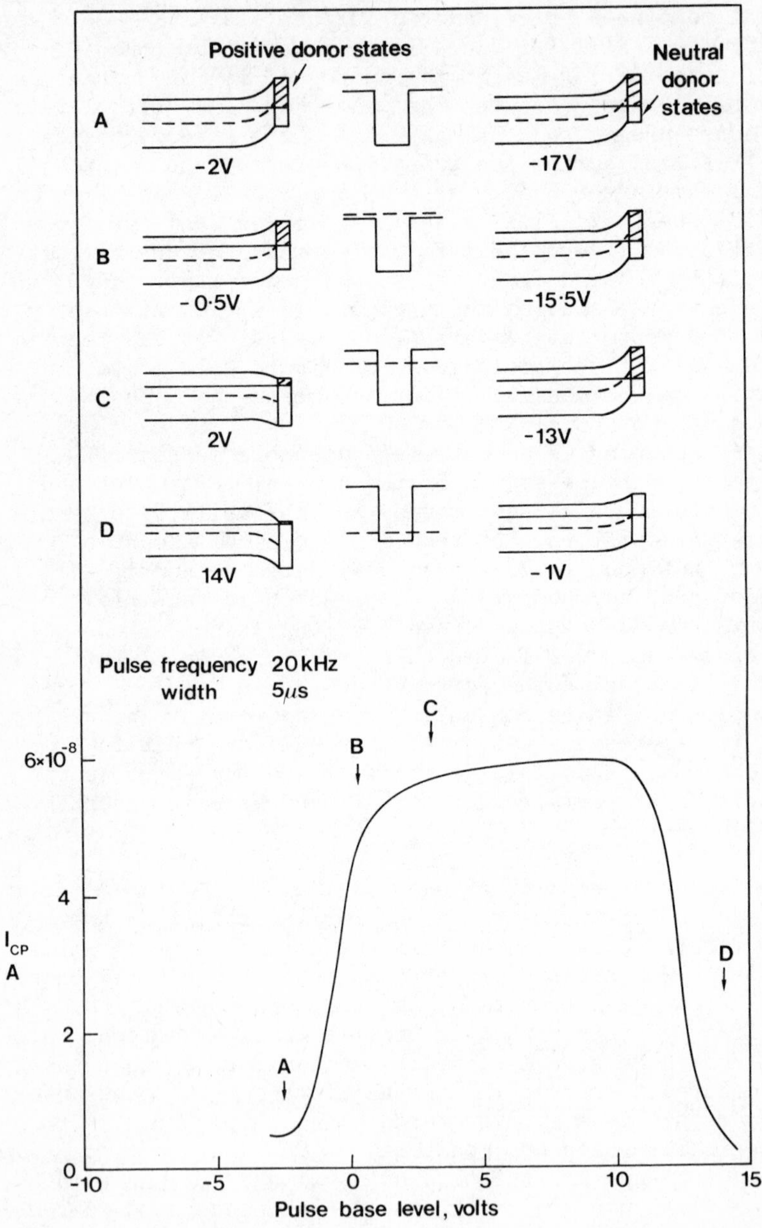

Figure 2.10 Charge pumping current as a function of gate pulse base level, together with schematic band diagrams at various points. (After Elliot [50]).

effects are minimized. Also, Reynolds [52] has shown that it is important that the measurement sequence and timings should be exactly the same at each measurement occasion; this is especially true if more than one instability is present, and if the act of measurement also constitutes a stress.

Sequin and Baldinger [53] have attempted to use the dynamic characteristics of MOS transistors to measure surface state densities resulting from exposure to γ-radiation. Their methods are based on the fact that when the transistor is turned "on" by a rectangular gate pulse, the drain current will initially overshoot its equilibrium value, decaying back as the surface states change their occupancy. Problems arise, however, because the channel conductance can only be measured in the inversion region, so that other parts of the bandgap can only be measured by use of a sampling pulse. Also, the very long time constants that they observed at room temperature raise some doubts as to whether the observed effects are due to surface states, or to some other mechanism, such as electron/hole trapping. As the method is relatively insensitive when the surface state density is low, it is not surprising that it is little used.

The surface state density can also be determined from the channel current of an unsaturated transistor operating in the subthreshold or weak inversion region. Although the value of D_{it} obtained from $I_D - V_G$ measurements can be effected quite strongly by the statistical variations in surface potential [54], van Overstraeten et al [55] have shown that the $I_D - V_D$ characteristics are insensitive to these fluctuations. This second method also has the advantage that the region of the bandgap between ψ_F and $2\psi_F$ can be investigated simply by altering the gate voltage.

2.5 Triangular Voltage Sweep Method

The measurement of mobile ion concentration was originally performed by monitoring the drift of the C-V curve along the voltage axis when the device was subjected to positive bias at high temperature. Since the drift saturates when all the charge has reached the Si/SiO_2 interface, the total mobile ion concentration may be calculated from the total voltage drift and the device capacitance. A faster and more sensitive technique, useful for process control purposes, has been described by Chou [56], and involves measuring the device current when a triangular voltage is applied to the gate of an MOS capacitor. The rate of change of voltage must be slow enough for the system to be in quasi-equilibrium if the resultant I-V characteristics are to be easily interpreted.

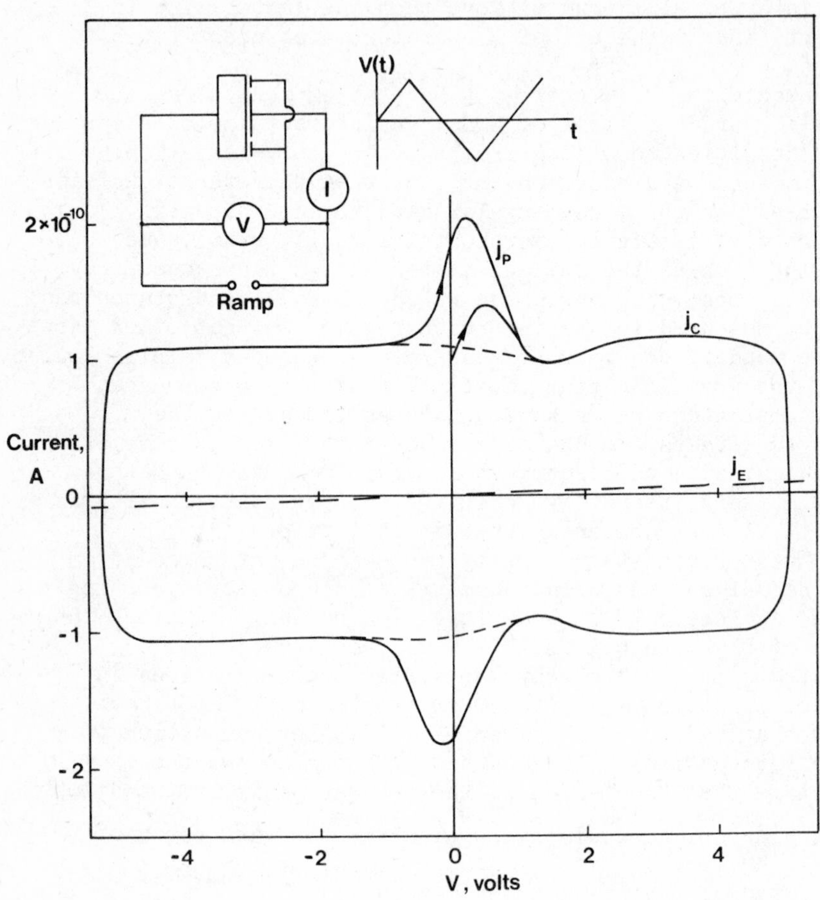

Figure 2.11 Typical I-V characteristic obtained by the triangular voltage sweep method. j_e is the electronic conduction current, j_p is the mobile ion current and j_c is the MOS capacitor charging current. (After Chou [56]).

This may be achieved by performing the measurements at elevated temperature, and by using a very slow voltage waveform. Practical limits are imposed on both these parameters by chemical reaction between the SiO_2 and the electrodes at high temperatures, and by the background current noise at low sweep rates, respectively. At typical experimental values of 240°C and 100 mV/sec, I-V curves of the form of Figure 2.11 are observed. Under these conditions the total current is made up of three components (other components may exist at higher temperatures, or, with systems other than the polysilicon-SiO_2-Si used by Chou, due to Faradaic currents). The normal steady-state electronic conduction current j_e is relatively small at the fields employed for these measurements, even at elevated temperatures, and will be roughly symmetrical (depending on the electrodes, since the conduction mechanism is interface limited). j_c is the displacement current due to the charging of the MOS capacitor. For an ideal parallel plate capacitor this component would simply lead to a rectangular I-V curve, the deviations at small voltages reflecting the changes in surface potential, since

$$j_c = C_{ox} \frac{dV}{dt} (1 - \partial\psi_s/\partial V) \qquad (2.11)$$

if dV/dt is small enough for the space charge and surface state population to follow the applied voltage (∂Q_s and ∂Q_{it} vanishingly small). This component of the current is thus that obtained by Kuhn [41] in the quasistatic C-V method. The third component, j_p, is the displacement current due to the motion of mobile ions. Under quasi-equilibrium conditions, this component is given by

$$j_p = - \frac{dV}{dt} \frac{Q_m}{d} \frac{d\bar{X}}{dV} \qquad (2.12)$$

where Q_m is the total mobile charge and $\bar{X}$ is the position of the charge centroid.

For a completely symmetrical electrochemical cell the displacement peaks occur at zero applied volts, but for practical MOS systems the differences in the electrodes are reflected in a slight asymmetry of the peaks. The total mobile ion density can be obtained from the area under the j_p peak, but information on the charge distribution can only be extracted by fitting a theoretical model to the observed curves. This, together with the problems of ensuring quasi-equilibrium without introducing other current components, results in the method being more useful for process control purposes than for a full investigation of the mobile ion transport mechanism.

2.6 Thermally Stimulated Ionic Conductivity (TSIC) Method

Instead of monitoring the MOS capacitor current at high temperature as the applied voltage is swept (as in the TVS method), the mobile ion may be characterized by measuring the

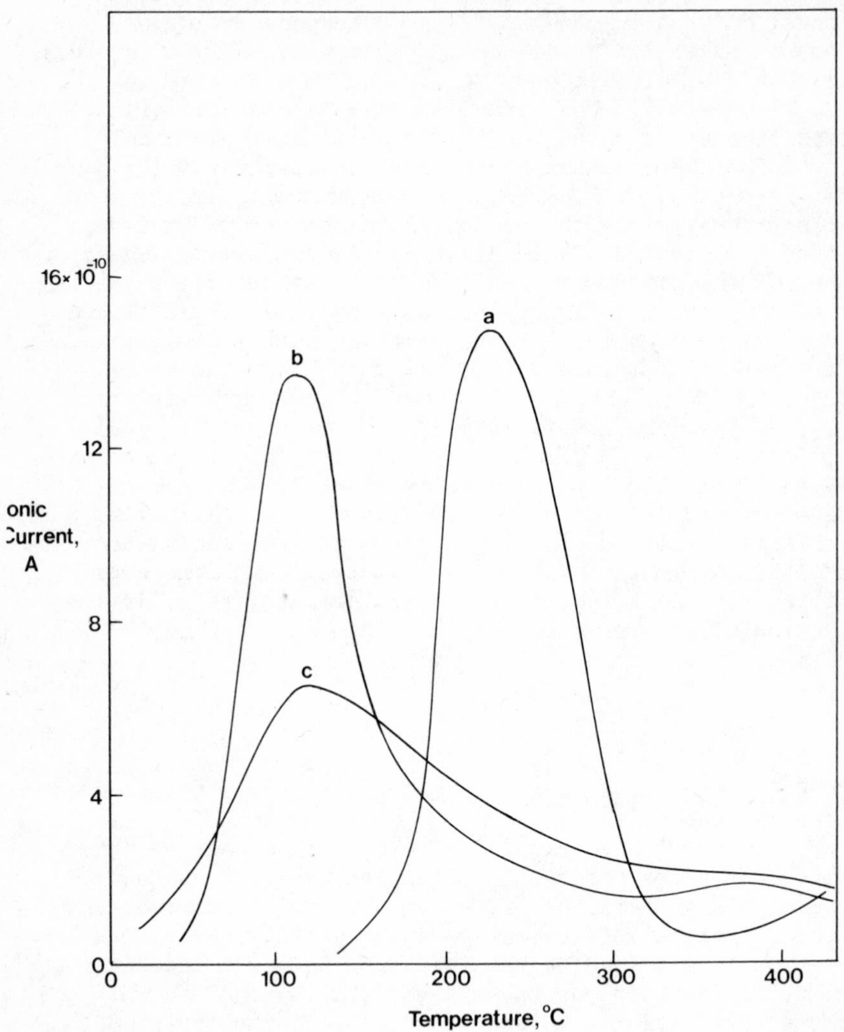

Figure 2.12 Successive thermally stimulated ionic current scan with hyperbolic heating rate for a sample with 3 x 10^{13} Na^+/cm^2 (a) first heating, aluminium positive, (b) second heating, aluminium negative, (c) third heating, aluminium positive. (After Hickmott [57]).

current as the temperature is increased at a fixed bias. (The TSIC method has also been used [57], with different heating and biasing sequences to suppress the effects of mobile ions and surface states, to study defect centres in the bandgap of the semiconductor).

Experimentally, the procedure consists of taking a thermal scan under positive bias up to a predetermined maximum temperature, cooling under bias, and then measuring a background curve for reference purposes with a second heating cycle (assuming no further ion motion). This procedure is then repeated with negative bias to determine the characteristics of ion motion from the silicon to the metal interface.

Hickmott [58] originally used a hyperbolic heating rate ($1/T = 1/T_o - at$), which simplifies the data analysis when using a simple single-trap model, but most authors have used a linear [58a] or an exponential [59] temperature ramp, and employed curve fitting techniques to extract the parameter values. At high enough temperatures, it is assumed that the bulk of the oxide is transparent to the ions (if the mobility is such that the transport from one interface to the other is fast compared to the temperature ramp rate), and that the rate-controlling step is the release of the ions from traps of depth ϕ. Using first-order kinetics, if n_t is the number of trapped charges, then

$$\frac{dn_t(t)}{dt} = -n_t(t)s\ \exp(-\phi/kT(t)) \qquad (2.13)$$

where s is the "attempt-to-escape" frequency. The capacitor current I(t) is simply given by

$$I(t) = -qA\frac{dn_t(t)}{dt} \qquad (2.14)$$

where A is the capacitor area. At $t = 0$, $n_t(0) = N_m$, the total number of ions present, which can thus be obtained by integrating the total ionic current.

For practical MOS systems, the single trap depth model does not give a complete description of the ion motion. For example, there may be more than one ionic species, or there may be a range of trap depths. In these cases Equation 2.13 applies to the change in occupancy of a particular trap, and the total ionic current is obtained by integrating Equation 2.14 over all possible trap depths.

Although requiring considerably more equipment and analysis than other methods of studying ionic motion (such as C-V shifts), the TSIC method is capable of providing a much fuller insight of the actual mechanisms of ion transport and trapping.

2.7 Thermally Stimulated Surface Potential (TSSP) Method

Another means of measuring the density and distribution of surface states in an MOS capacitor is the thermally stimulated surface potential method described by Yamashita et al [60]. In this method, the states are first filled by the application of gate bias at room temperature and then the sample is cooled to liquid nitrogen temperature to freeze the captured charge. Next, a non-equilibrium charge is induced in the semiconductor by the momentary application of a different gate bias, and finally the surface potential of the open-circuited gate is monitored as the sample is heated back towards room temperature. By repeating this procedure with different quantities of non-equilibrium charge, and with knowledge of the point at which the steady-state condition is regained, the surface state density can be calculated over a wide range of the bandgap. TSSP is a relatively new technique, and has not yet been widely used.

2.8 Optical Methods

The investigation of the trapping of holes or electrons at trap sites within the insulator forbidden gap is frequently performed by optical methods, using MOS transistors or capacitors with semi-transparent electrodes (100Å of aluminium is typical). In contrast to the well defined methods described earlier for the investigation of ionic and surface state instabilities, nearly all the workers using optical techniques have their own preferred method, and hence only a general indication of the methods will be given here.

The methods generally revolve around the measurement of photo-currents as the traps are either emptied or filled. Optical photons may create electron-hole pairs in the semiconductor directly, and these carriers can be made to drift towards the interface by the application of a suitable field. If the electrons gain sufficient energy they can be injected across the interface barrier into the oxide, where most will flow to the top electrode; a small percentage will be trapped, however. This method has been extensively employed by Ning and his co-workers [61, 62, 63].

By use of vacuum ultraviolet (VUV) radiation, carriers may be introduced directly into the insulator. Powell [64], amongst others, has used this method, and has varied the position of the trapped charge (holes) by changing the penetration depth of the radiation (dependent on the photon energy). Since the position of the charge distorts the field in the insulator, he has also recorded changes in electron tunnel injection at high fields resulting from hole trapping. Thus photo I-V measurements can give a value of the charge centroid, which is not obtainable with the C-V technique. Woods and Williams [65] have filled hole traps by a corona discharge near to the surface of a bare oxide, a method which necessitates the use of a mercury electrode, or the re-evaporation of the electrode after charge trapping has taken place. They then determine the position of the trapped charge by successive etchings of the oxide surface.

The chief disadvantage of all the methods described is that they lead to very high densities of trapped charge if the injection is continued until the steady-state condition is reached, so that the magnitude of the flat-band voltage is increased dramatically, above the oxide dielectric strength in some instances. Only a small percentage of this charge can subsequently be detrapped by photodepopulation techniques. Thus it is not always clear whether the photo-currents observed result from the same trapping centres which are filled during service-type bias-temperature stress.

One of the most sophisticated and all embracing optical techniques is that reported by Kapoor et al [66] to investigate electron trapping centres. The first step is current monitored photoinjection; electrons are introduced into the oxide by illumination from a deuterium lamp, with a positive field applied (positive voltage to the outer, transparent electrode). This step is continued until the current saturates, indicating that a steady state trap population has been reached. The main part of experiment then consists of spectrally resolved photodepopulation measurements. Photocurrent-vs-wavelength spectra are obtained as the wavelength is swept from 700 to 300 nm (ie increasing photon energy) with a fixed applied bias. Two spectra are taken, one with a fast sweep rate of wavelength and a low illumination intensity, so that the trap occupation is not significantly altered by the measurement, and the other with a slow sweep rate and high intensity, which results in the optically accessible traps being emptied by the sweep. Typical spectra for these two limiting cases are shown in Figure 2.13. These results are then normalized to account for the output characteristics of the monochromator and the internal reflections in the oxide, allowing the spectral response (defined as the number of electrons released per unit electrode area per unit photon-energy interval, divided by the number of photons per unit volume in the oxide) $R(h\nu)$ to be plotted:

$$R(h\nu) = \frac{I/(qAr)}{(I/h\nu)(u/s)(s)} \quad \text{cm eV}^{-1}. \qquad (2.15)$$

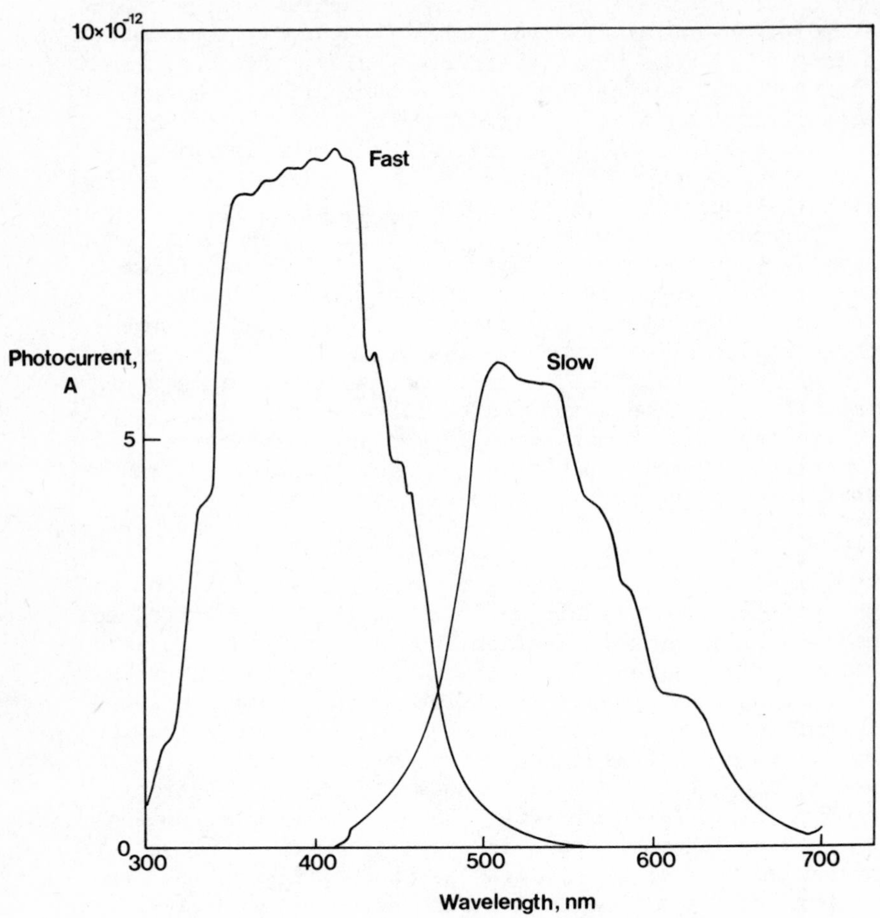

Figure 2.13 Typical spectrally resolved photodepopulation characteristics, as the light is swept from long to short wavelengths. (After Kapoor et al [66]).

where I is the photocurrent, A is the electrode area, r is the sweep rate, u/s is the total energy density in the oxide per unit incident light intensity, and s is the incident light intensity at the top electrode. Figure 2.14 shows the spectral responses for the fast and slow-sweep limits, for three oxide thicknesses. From the fast-sweep-limit spectral response the trap depth E_T is given as the intercept on the x-axis of the high photon energy part of the curves, the

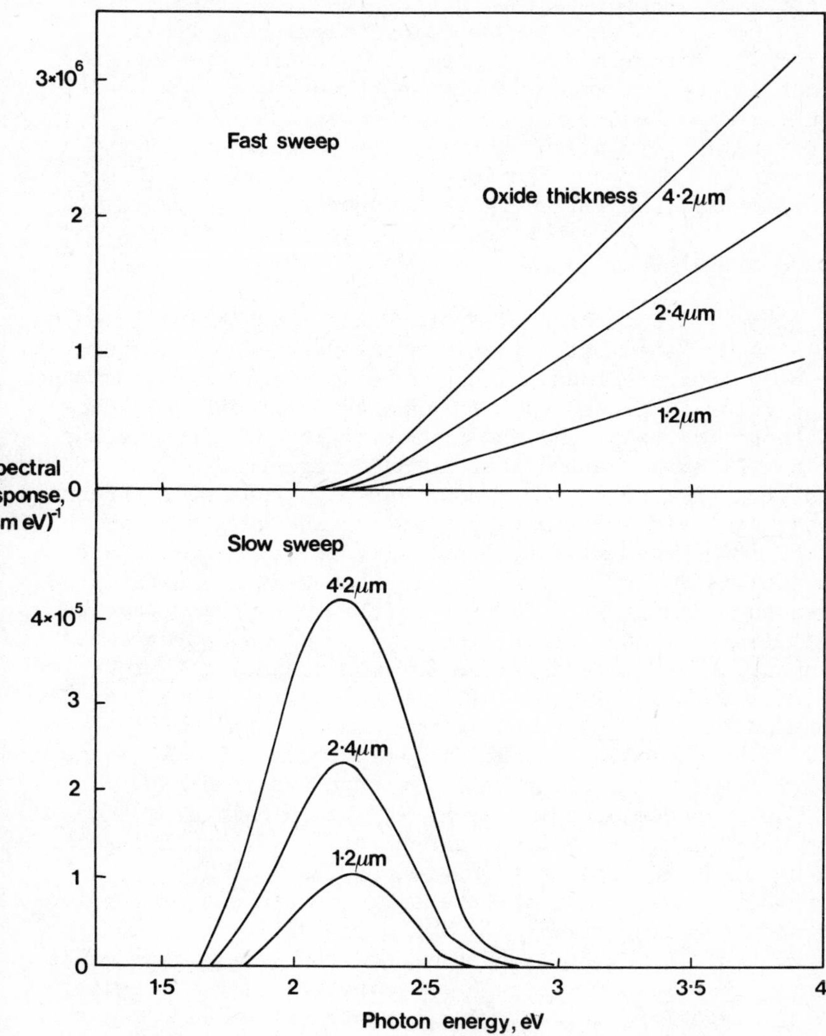

Figure 2.14 Spectral responses for three thickness of oxide film for the slow and fast sweep-rate limits. (After Kapoor et al [66])

actual distribution of trap energies being taken from the bell-shaped curve of the slow-sweep limit response. The total amount of detrapped charge may be obtained from integrating the photocurrent during the slow-sweep scan.

Methods such as those described above have yielded important information on the trapping centres present in thermal silicon dioxide. There is, however, a lack of data on which traps become filled when MOS transistors are subjected to their normal working conditions, and of the mechanisms of charge injection in this situation. This gap could, perhaps, be filled by photodepopulation experiments on devices which have been subjected to long-term bias temperature stress.

2.9 Summary: Selection of Appropriate Measurement Techniques

Although the most commonly used experimental techniques have been described many minor variations exist, so that the investigator is presented with an often bewildering choice of methods for examining the instabilities of a particular device. It is suggested that low-and high-frequency C-V plots, together with a dc measurement of transistor threshold voltage and gain, will give an initial indication of which instabilities are present, especially if measurements are made both before and after positive and negative bias-temperature treatment. Further information on any particular instability can then be obtained by use of an appropriate technique. For surface states the conductance technique is still the most widely used, although for small area transistors charge pumping is suggested as the simplest and most sensitive technique, capable of giving both "global" state densities and their distribution within the band-gap.

The presence of mobile ions may be quickly detected using the triangular voltage sweep method, although care must be taken in interpreting the results. If it is required to investigate the transport mechanisms of the ions, then the thermally stimulated ionic conduction method is the most suitable. Selection of a method for investigating charge trapping (particularly slow hole-trapping) presents rather more of a problem. Direct measurements of device parameters under accelerated ageing may be extrapolated to service conditions, but the details of the mechanism cannot be easily extracted from the methods normally used to examine charge trapping, namely optical methods.

3

Surface States

3.1 Introduction

The concept of surface states has already been introduced in Chapter 1. The termination of the periodic semiconductor lattice results in the existence of localized energy states within the bandgap, and which can be charged or neutral depending on the position of the Fermi level. The states reside at the semiconductor/insulator interface (although, historically, the term "slow states" was sometimes used for the states arising at the insulator/ambient interface of a semiconductor with only a thin, native oxide and exposed to the atmosphere) and may be regarded as an intrinsic part of the MOS system. The states may be either donors (positive when charged) or acceptors (negative when charged), and are filled according to the distribution functions [8]:

$$F_{SD}(E_T) = \frac{1}{1 + g \exp\left(\frac{E_F - E_T}{kT}\right)} \quad \text{(donors)} \qquad (3.1)$$

and $$F_{SA}(E_T) = \frac{1}{1 + 1/g \exp\left(\frac{E_T - E_F}{kT}\right)} \quad \text{(acceptors)} \qquad (3.2)$$

where E_T is the energy level of the state, E_F is the Fermi level and g is the ground state degeneracy, which in silicon is 2 for donors and 4 for acceptors. For practical purposes, however, at room temperature and in equilibrium, donor states above the Fermi level can be regarded as positively charged with those below it neutral, whilst acceptor states are negatively charged when below the Fermi level, and neutral when above it. In fact, the distinction between acceptors and donors is often not made explicit, because from a device viewpoint it is the change in occupancy when the surface potential is altered that is of importance. Also, it has been shown by Hughes [67] that the distinction cannot be made from the results of many experimental techniques (including C-V methods) when a fixed oxide charge is also present. By contrast, Ziegler [68] has used the conductance technique and, based on the assumption of some negative oxide charge, has stated that the states are principally donors, whilst Schulz and Johnson [69], using DLTS, have found only acceptors.

The rate at which surface states empty and fill when the surface potential is suddenly changed may be described by a time constant τ, which is a strong function of the energy level of the state. The variation of τ with the surface potential ψ_s, as measured in 1967 on samples with fairly high surface state densities ($> 10^{11}$ states/eV/cm^2) by Nicollian and Goetzberger [10] is shown in Figure 3.1. These curves can be fitted by the expressions

$$\tau = \frac{1}{\bar{\nu}\sigma_p n_i} \exp\left[\frac{-q(\psi_B - \psi_s)}{kT}\right] \quad \text{for p-type} \qquad (3.3)$$

$$\text{and} \quad \tau = \frac{1}{\bar{\nu}\sigma_n n_i} \exp\left[\frac{q(\psi_B - \psi_s)}{kT}\right] \quad \text{for n-type} \qquad (3.4)$$

where $\bar{\nu}$ is the mean thermal velocity and σ_p and σ_n are the capture cross-sections for holes and electrons respectively (Fahrner and Goetzberger [70] have used a similar expression for the time constant of a single state introduced by ion implantation). Taking $\bar{\nu} = 10^7$ cm/sec and $n_i = 1.6 \times 10^{10}$ cm^{-3} for silicon at room temperature, these results (for steam-grown oxides) give $\sigma_p = 2.2 \times 10^{-16}$ cm^2 and $\sigma_n = 5.9 \times 10^{-16}$ cm^2. These values are independent of the energy level of the state, but do depend on the oxidation conditions. Later results by Deuling et al [71] taken on n-type silicon with dry oxides give mid-gap values of $\sigma_n = 10^{-14}$ and 10^{-15} cm^2 for (100) and (111) substrates respectively. Near the conduction

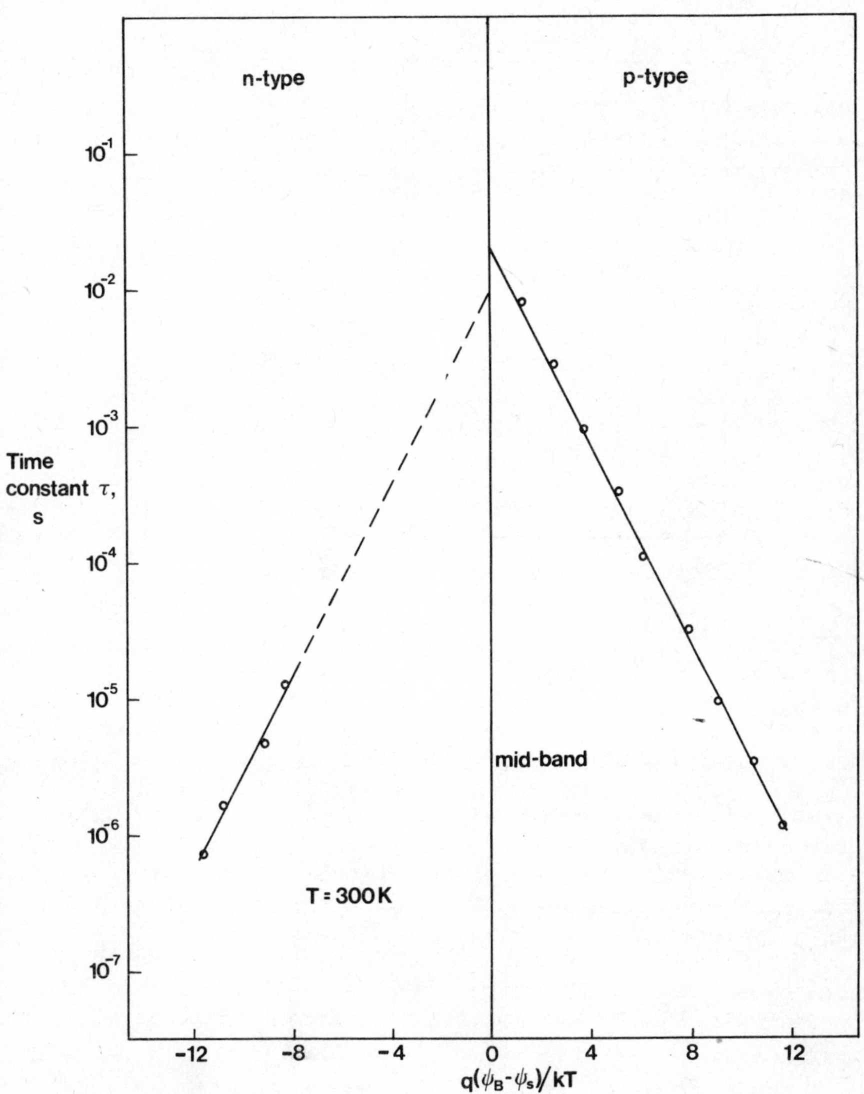

Figure 3.1 Variation of surface state time constant with surface potential as measured by Nicollian and Goetzberger [10], using the conductance technique.

c

band edge, however, they found an exponential decrease in σ_n with energy, as is shown in Figure 3.2. By extending the frequency range of the conductance technique to 100 MHz,

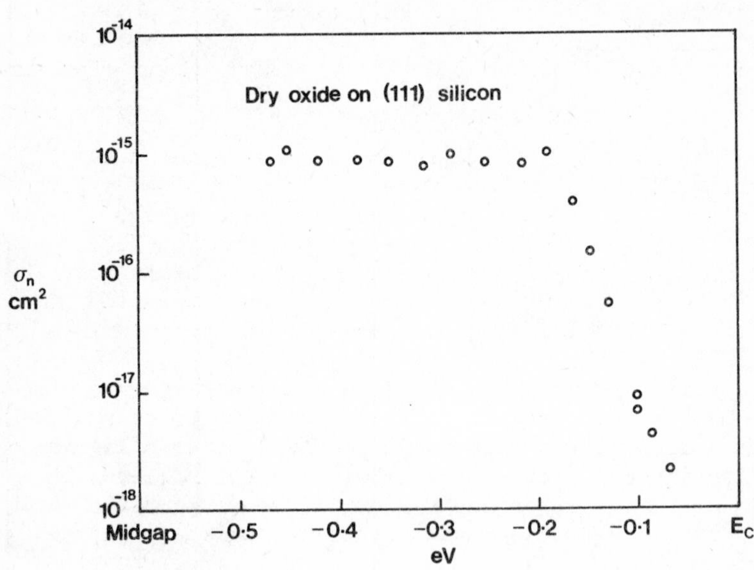

Figure 3.2 Variation of electron capture cross-section of surface states with energy level. (After Deuling et al [71]).

Morita et al [72] have reproduced Deuling's results, but also found a second (high-frequency) peak in the (G_p/ω) - log (f) curve, which they tentatively ascribe to previously unknown states near the band-edge.

The calculation of capture cross-sections from conductance data has been shown by Muls et al [73] to be strongly dependent on the assumed distribution of surface potential fluctuations (discussed in Section 3.4) which to some extent accounts for the wide range of values reported. For example, Schulz and Johnson [69] found a hole capture cross-section of 5×10^{-13} cm^2 in high-quality oxides, and Wang and Evwaraye [46] reported a value of 1.6×10^{-14} cm^2 for the discrete level state introduced by gold implantation.

3.2 The Origins of Surface States

It is well known that thermal oxidation of silicon can reduce the density of surface states from approximately one state per surface atom to less than 10^{12} states/cm^2, but that defects in the interface prevent complete passivation. It is not clear, however, what form these defects take. Laughlin et al [74] have made a theoretical study of various types of possible defects, and found that there are no surface states at all for an ideal interface with no dangling bonds or bond-angle distortion. This latter defect introduces a tail of states from the conduction band edge, as observed experimentally in oxides with $N_{it} < 10^{10}$ states/eV/cm^2 by Schulz and Johnson [69]. The presence of Si-Si bonds and dangling Si bonds in the oxide both give rise to a tail of states from the conduction and valence band edges, a feature often observed experimentally (see Section 3.3). Finally, the only defect type which produces a state near mid-gap is a dangling Si bond on the semiconductor side of the interface, that is, an oxygen vacancy in the interface region.

In a slightly different model of the interface, Svensson [36] also considers surface states to be caused by trivalent silicon atoms with bonds to three other silicons. He also considers two other forms of trivalent silicon, in the oxide or in the interface region, to be responsible for the hole traps (the subject of Chapter 6) and the positive fixed oxide charge, Q_f. Thus in this model, the strong relationship between surface state density and the total interfacial charge (Q_f + trapped holes), both initially and after stressing, is attributed to a common chemical origin rather than to a coulombic interaction between positive charge centres in the oxide and the semiconductor space-charge region, as has been suggested [8].

The states discussed above are often regarded as an intrinsic property of oxidized silicon surfaces. Additional states are often observed when additional charge is created at the interface by, for example, drifting mobile ions through the oxide, or exposure to ionizing radiation. Once again, these states could either result from the coulombic action of the charge centre, or from structural changes at the interface resulting from their incorporation.

3.3 Surface State Distributions

One of the most important parameters of surface states, both from the practical viewpoint of their effect on device characteristics, and from a theoretical consideration of the physics of the solid-solid interface, is their distribution in the forbidden band. Early measurements (1966) were made by Gray and Brown [43], using the temperature technique described in Section 2.1.4. The measurements were made in the regions near E_C and E_V using capacitors on n- and p-type silicon respectively, and showed substantial peaks at ~0.1 eV below the conduction band and above the valence band, with a broad minimum in midband. With the relatively primitive oxidation techniques employed, very high values of D_{it} (approaching 10^{14} states/cm^2/eV) were found. However the assumptions made in the temperature technique have subsequently been challenged, throwing severe doubt on the existence of peaks in the distribution near the band edges. In particular, the validity of the assumption that the distribution of the states, and their properties, are temperature independent between liquid nitrogen and room temperatures is by no means obvious. Also, at the low temperature required to investigate the regions near the band edges, the requirement for the measuring signal to cause the surface potential to sweep only a fraction of kT/q is no longer met. Lam [75] has measured the state density of a single sample over the majority of the band gap using a surface photovoltage technique, and his results, displaying a broad minimum without peaks, are shown in Fig 3.3, which also shows the distribution of the same sample measured with the temperature techniques for comparison. Note the overall improvement in state densities in this case compared to those of the original Gray-Brown paper. An absence of peaks has also been reported by Deuling et al [71]. They used Nicollian and Goetzberger's [10] conductance technique with dry oxides on (111) substrates and found a much shallower distribution than Lam, as is shown in Fig 3.4, where it should be noted that the measurements have been taken at low temperatures.

Although it is now generally accepted that the density of surface state increases steadily at the band edges, Arnold [76] found that they reached a plateau at about 0.05 eV from each band edge by using a method based on the channel conductance of an MOST in inversion (and hence exploring the opposite half of the forbidden gap to the capactiance techniques). However, although the disagreement of the various methods on the shape of the surface state distributions near the band edges makes a physical description of the exact causes of surface states difficult, it is not of too great a concern when considering device applications. In this latter case it is the states near mid-gap, which can change their occupancy during device operation, which need to be measured accurately. For new (unstressed) devices of high quality, the distribution is widely held to be a broad minimum, although after stressing structure often develops near mid-gap.

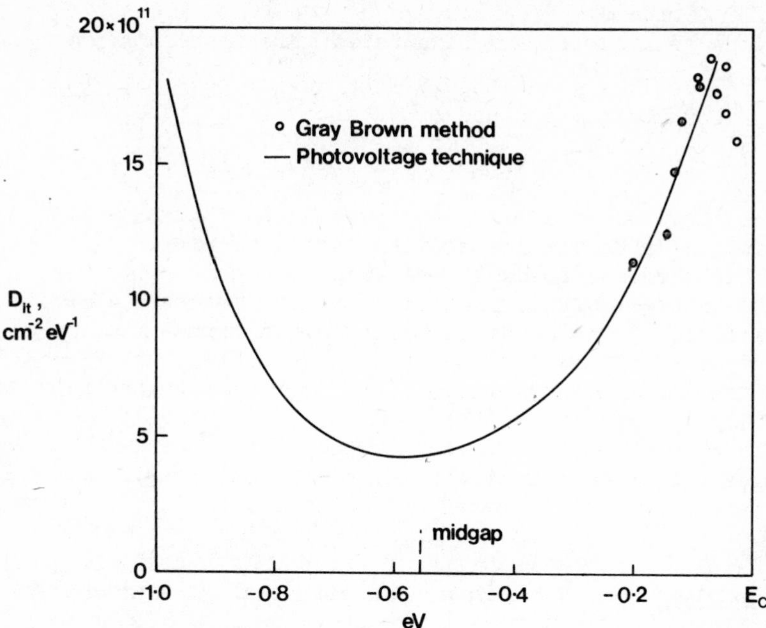

Figure 3.3 Variation of surface state density with energy (After Lam [75]).

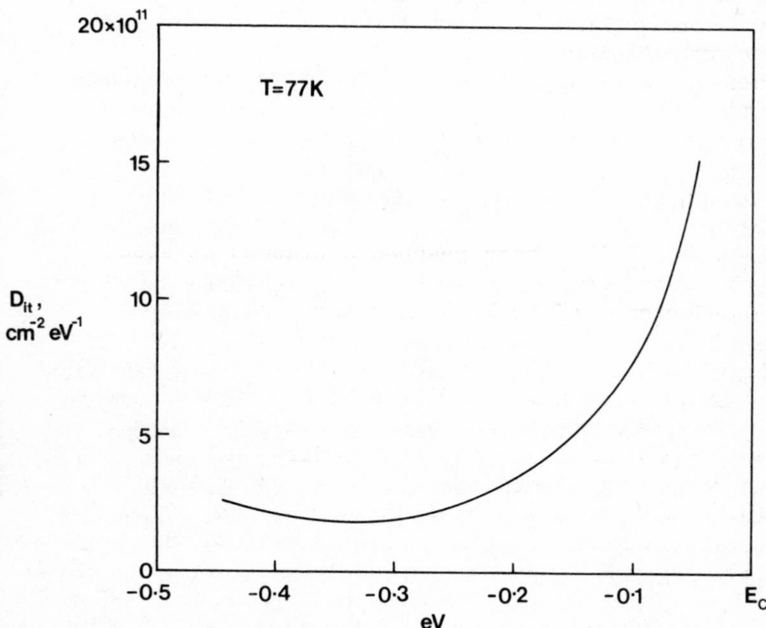

Figure 3.4 Variation of surface state density with energy (After Deuling et al [71]).

3.4 Effect of Lateral Nonuniformities of Oxide Charge on Surface State Distributions

In Section 2.2 the broadening of the conductance peak from its ideal value was modelled by assuming a random fluctuation of the fixed charge density. In Nicollian and Goetzberger's original model [10], the charges were considered to produce a Gaussian spread of the surface potential, and although Brews [77] produced a more sophisticated model, it has since been shown [73] that for small fluctuations of the surface potential the models give similar values of the surface state density (assumed to be laterally uniform).

According to the Brews model, the standard deviation of the surface potential, σ_s^2, should be proportional to the mean fixed oxide charge $\overline{Q}_f$. This relationship was not observed experimentally from conductance-against-frequency measurements by Werner et al [48], who found instead that σ_s^2 was related to the initial value of the surface state density. They also found that σ_s^2 increased in proportion to the number of mobile ions drifted to the interface, and interpreted these results as implying that the charge centroid of the fixed charge is located up to 90Å into the oxide, and that only charges very near to the interface induce surface states.

In the preceeding discussion the variations of surface potential considered have been small, and have been derived indirectly from fitting theoretical models to conductance data. More recently, Zamani and Maserjian [47] have developed an experimental method for determining the distribution of surface potential by making non-equilibrium C-V measurements at low temperature (so that there is no contribution to the capacitance from states changing their occupancy). They found that, after being subjected to irradiation treatment, their samples exhibited a very wide spread (>15V) of the flatband voltage distribution density. On the basis of room temperature C-V curves, these samples would traditionally have been characterized by a large increase in the mid-band surface state density, caused by the irradiation. The presence of such large lateral nonuniformities of charge, if confirmed, raises serious questions regarding the proper interpretation on the apparent phenomena of surface state growth during stressing, to be described in Section 3.6.

3.5 Dependence of Surface State Density on Processing

There has been a large reduction in the density of surface states from the earliest oxides to those of the present day; indeed, much effort was originally made to reduce the density to a level which allowed useful devices to be made. The changes of processing technology which have resulted in these reductions have been fairly minor, consisting mainly of optimizing such factors as the oxidation temperature and ambient and particularly, the annealing treatment. Because of poorly characterized factors such as preoxidation cleaning, there is incomplete agreement on the relationship between D_{it} and all the processing variables, but some general dependencies are described by Razouk and Deal (1979) [22].

The often observed correlation between D_{it} and the fixed oxide charge Q_f was mentioned in Section 3.2. Thus, for dry oxides D_{it} increases with decreasing oxidation temperature (Deal's famous "oxygen triangle" [78]), and as the silicon orientation is changed from (100) to (110) to (111). The temperature dependence can be explained by recalling that oxide growth takes place by the inward diffusion of oxygen, so that at low temperatures, where the oxygen mobility is low, a large number of oxygen vacancies are formed at the interface, resulting in the defects described by Svensson [36]. The temperature dependence is less clearly understood for wet oxides, or for annealed dry ones. For example, Wang [79] has reported that for high quality oxides D_{it} is reduced (to $< 10^{10}$ states/cm^2/eV) by cooling the samples very slowly after oxidation. This is in contradiction to the dependence on pull-rate from the hot zone of the furnace reported by Deal [78], and is explained by stating that, when the effect of oxygen vacancies is removed by annealing, any residual states are caused by "quenched-in" interstitial (active) impurities. Other authors [80] have considered the effect of strain resulting from the mismatch of the coefficients of thermal expansion.

Wet-grown oxides generally have a lower surface state density than equivalent dry-grown ones, leading Balk to postulate that a hydrogen species (H or OH) forms a bond with the silicon at the oxygen vacancies, and thus satisfies the dangling bonds. Deal, Mackenna and Castro [81], and Castro and Deal [82] have studied the annealing of surface states in nitrogen and forming gas (H_2/N_2) at low temperatures (300-600°C), and found reductions with forming gas at all temperatures and also, very significantly, for both forming gas and pure nitrogen if the annealing was performed after the deposition of an aluminium field plate. This suggests that the aluminium reacts with the small quantity of water present at the metal-oxide interface to form active hydrogen ions which migrate to the Si-SiO_2 interface, where they anhilate the surface states. This is supported by the fact that less reactive metals than aluminium do not produce such effective annealing, and also by the reduction in annealing properties caused by an intervening layer of high-density silicon nitrode. For short-channel aluminium gate transistors it is possible that the annealing effect will be reduced because the structure can no longer be regarded as essentially two dimensional, and active hydrogen may be lost by sideways diffusion. This effect has indeed been found by Schlegel [83].

For silicon gate transistors, production of active hydrogen by the reaction of water with aluminium is no longer possible in the channel region. Also, there are now two interfaces to be annealed, since surface states at the (degenerately doped) poly-silicon interface can lead to charge injection problems. Hickmott [84] has studied the annealing of these structures in detail. He once again found that no annealing occurs from a purely thermally activated reaction, and that the annealing temperature needs to be high enough to allow hydrogen to permeate into the SiO_2. Above 500°C, however, he postulates that a second reaction takes place which results in an increase in surface states and fixed charge:

$$\underset{\text{solid}}{Si} + \underset{\text{solid}}{SiO_2} \rightarrow \underset{\text{gas}}{2SiO}$$

the gaseous SiO condensing to form polymers of $(SiO)_{2,3....}$. At still higher temperatures (>700°C) there is again some reduction in surface state density, possibly due to the formation of SiOH.

The use of small additions (~1%) of HCl to the oxidizing ambient is discussed in some detail in Chapter 4, in connection with its property of reducing mobile ion concentrations. Severi and Soncini [85] have found that HCl oxides also exhibit significantly lower surface state densities than ordinary dry ones. The distributions of surface states for HCl and dry-oxygen oxides obtained by them with the quasistatic technique (near its resolution limit) are shown in Figure 3.5. The mechanism proposed for this reduction is again formation of Si-H bonds. Low temperature annealing is still required to obtain these very low densities, because the Si-H bond is not very stable

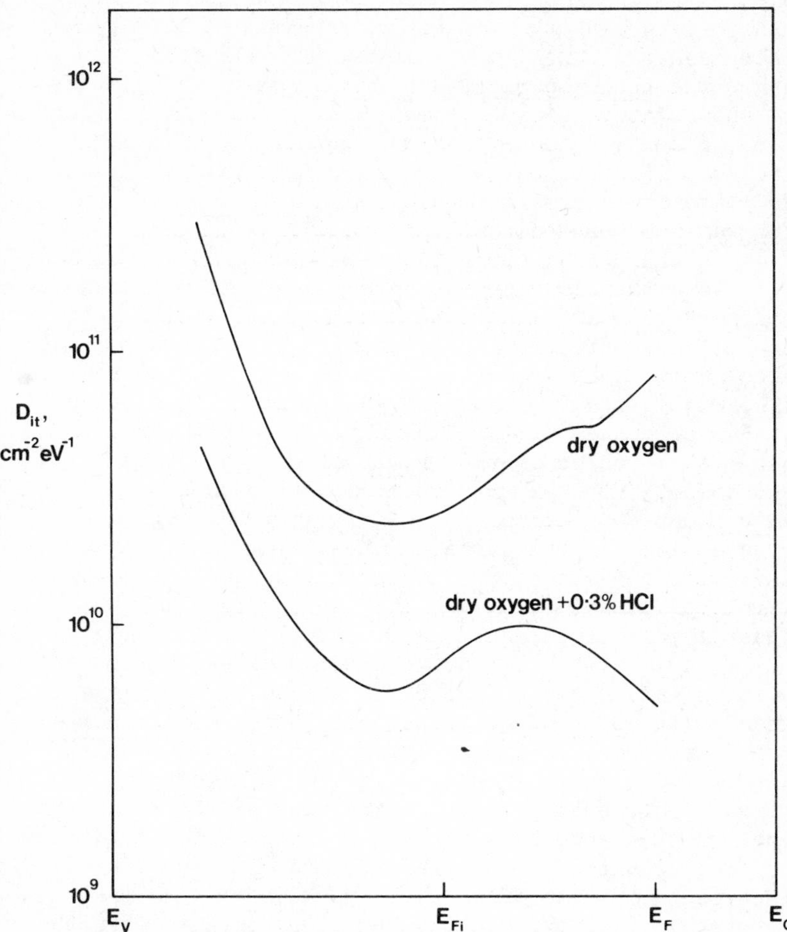

Figure 3.5 Reduction in surface state densities produced by additions of HCl to the oxidizing ambient; (100) silicon. (After Severi and Soncini [85]).

at the oxidation temperatures. Interestingly, they found that the reduction in surface states is not accompanied by a concomitant reduction in fixed charge, thus favouring the theory that they are caused by oxygen vacancies at and near the interface respectively, rather than by the positive charge inducing surface states by coulombic interaction.

3.6 Growth of Surface States During Stressing

Compared to the amount of effort spent on minimizing the surface state density of as-grown oxides, relatively little has been published on the increase in density with stressing. The three main categories of stress are positive and negative bias-temperature (B-T) treatment (where the polarity refers to the voltage applied to the gate electrode with respect to the grounded substrate), and exposure to various forms of radiation. Negative B-T stressing was used predominantly by the early workers, both because it is the normal operating condition of the p-channel MOS transistors then in wide use, and because it produces larger increases in surface state density than positive bias. Radiation effects were originally studied only by the military and space agencies, but with the introduction of electron beam lithography, ion implantation and plasma etching they are now also of concern to semiconductor manufacturers.

In all the studies of the growth of surface states published to date, the experimental results have been analyzed without considering the possibility of large spatial variations in the surface potential. Thus it may be necessary to re-assess the interpretations of these results in the light of Zamani and Maserjian's [47] discovery of large fluctuations of the surface potential.

One of the most comprehensive studies of surface state growth with negative B-T stressing is due to Goetzberger, Lopez and Strain [86]. They found that the states developed as a broad peak near mid-gap, although sometimes a double peak was observed, as is shown in Figure 3.6. The position of the (most prominent) peak was a function of the metallization used and, to a certain extent, on the oxidizing and annealing conditions. Saminadayer and Pfister [87] have also observed the growth of a double peak under negative B-T treatment for wet oxides which had been irradiated with 1 MeV electrons after growth. At relatively low stress levels (<175°C for 18 hours) there was a broad peak near mid-gap, but at higher temperatures, a much narrower peak at about E_V + 0.35 eV developed in both irradiated and non-irradiated samples. From these results it may be postulated that the broad peak is the result of positive charge build-up at the interface; when holes are trapped (see Chapter 6) near the interface there is a re-arrangement of the chemical bonding resulting in dangling Si bonds and hence surface states. This hypothesis thus explains the often observed result [51, 78] that the growth of surface states is directly proportional to the increase in positive interfacial charge. The narrow peak at E_V + 0.35 eV is probably caused by bond-breaking at the interface, the bonds

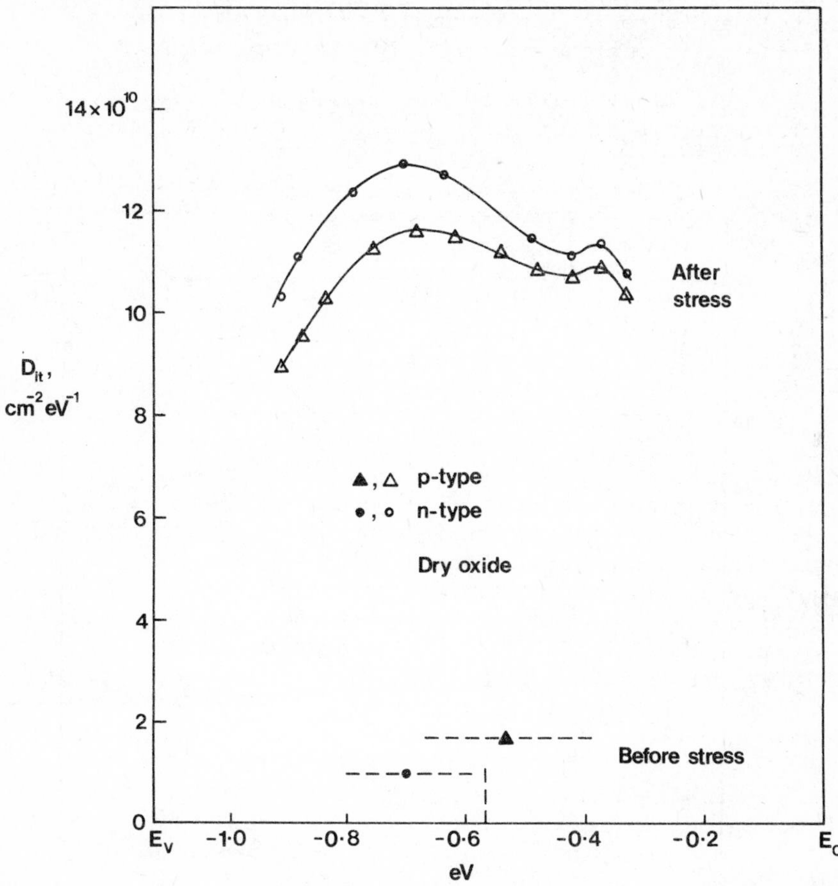

Figure 3.6 Growth of surface state density as a function of energy level. (After Goetzberger, Lopez and Strain [86]).

involved being either Si-O or Si-H. The results of annealing experiments [87] suggest that the latter is more likely.

Several other authors have observed the growth of surface states near mid-gap during negative B-T stressing. Semuskina and Semushkin [88] reported a sharp peak 0.17 eV above the band centre, whereas Elliot [50] placed a much broader peak near to mid-gap. Kasabov and Ilieva [89] found peaks 0.19 eV and 0.30 eV below the band centre, the number of states increasing by a factor of three after only 30 minutes at 200°C with an applied field of -1.3 MV/cm.

In addition to the growth of mid-gap states, some authors [88, 90] have observed a reduction of the states near the band edges. These results are thought, however, to be due to incomplete annealing of the samples prior to stressing.

The rate of growth of mid-gap states as a function of time and temperature are shown in Figure 3.7. It can be seen

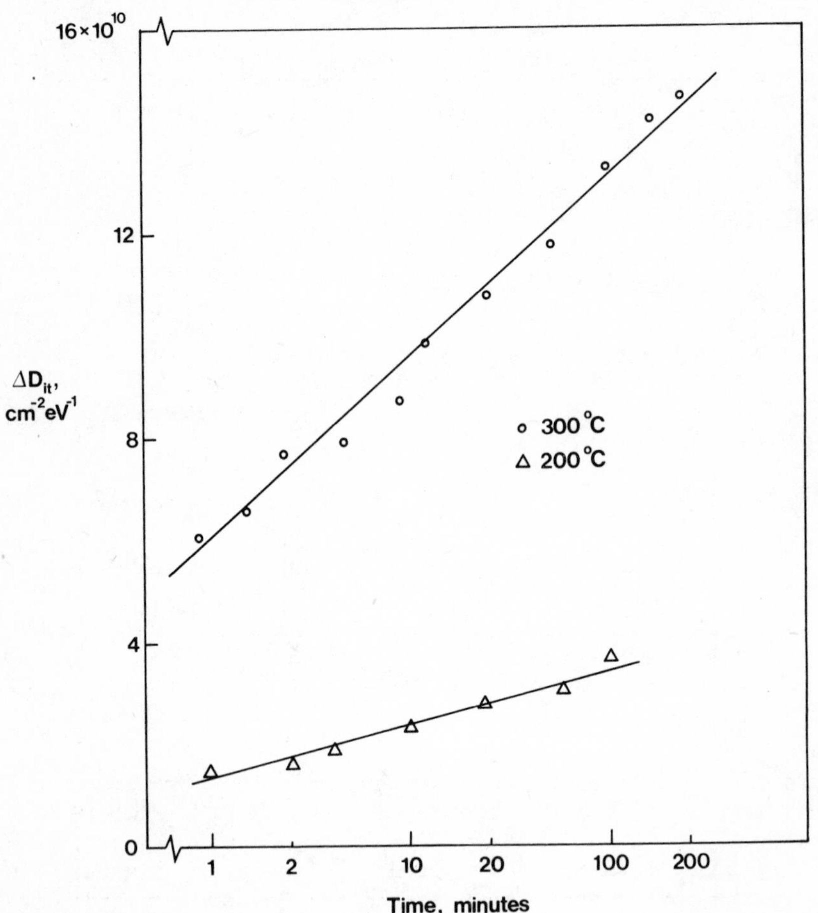

Figure 3.7 Growth of surface state density with time, with a field of -1 MV/cm, and with temperature as a parameter. (After Goetzberger, Lopez and Strain [86]).

that the density of states increases linearly with logarithmic time, and that the process has a strong temperature dependence;

an approximate activation energy of 1.4 eV can be obtained from these results. It has been observed [86] that the rate of growth is directly proportional to the initial surface state density, and it is also worth noting that the rate is generally much slower than other instability mechanisms, for example, mobile ion drift and dipolar polarization. Increases in the surface state density with negative B-T stressing similar to those in Figure 3.7 have been observed by the author with longer stress times and

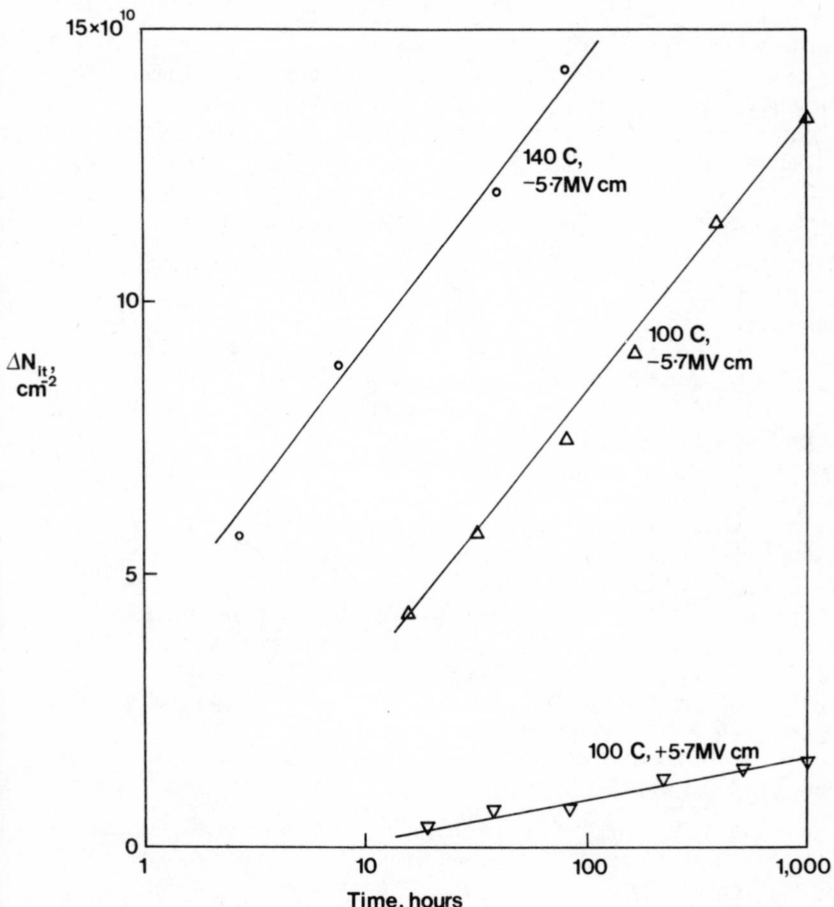

Figure 3.8 Increase of surface states with high field stressing for a p-channel MOS transistor fabricated with a commercial process. Measured with the charge pumping technique, for gate voltages sweeping from +1V to -14V.

lower temperatures, but with very high fields (~6 MV/cm). These results, shown in Figure 3.8, were obtained using the charge pumping technique on commercial p-channel, metal gate transistors and represent the total number of states that change their occupancy when the gate voltage is swept from +1 to -14 volts. With negative bias stressing the increase in surface states was closely mirrored by a change in the transistor threshold voltage which was attributable to the build-up of positive interfacial charge. The activation energy obtained in this study was 1.1 eV, very similar to the value often reported (see Chapter 6) for the slow hole-trapping mechanism.

Although the growth of surface states with negative B-T stressing is normally plotted against time on semi-logarithmic axes, Jeppson and Svensson [90a] have developed a model which predicts a $t^{\frac{1}{4}}$ time dependence. This model, in which equal numbers of surface states and trapped holes are formed, is based on the diffusion of a mobile species (probably an OH group) away from the Si/SiO_2 interface, and is discussed in more detail in Chapter 6. Figure 3.9 shows the results obtained by Jeppson and Svensson at two temperatures, and it can be seen that a $t^{\frac{1}{4}}$ dependence is observed except at the highest fields at room temperature; some criticism may be levelled at these results, however, because after each period of B-T stress the capacitors were brought back to room temperature with the gate floating, hence producing an unknown degree of annealing. At very high fields, the growth of surface states is linear with time, a behaviour also observed for other instabilities (for example, electron trapping) in conditions where the oxide is conducting heavily.

Figure 3.8 shows that there is a small growth in the surface state density with positive B-T stressing. In this case, the shift in the transistor threshold voltage saturated after a few hours, although the increase in surface states was accompanied by a continuing fall in the gain factor. Shiono et al [91] have also observed an increase in D_{it} without a concomitant change in the oxide charge with positive B-T stressing. In their experiments, performed on n-channel, poly-silicon gate transistors and using the quasistatic technique, the states were located 0.15 eV above mid-gap, and the activation energy of the process was 1.0 eV. The mechanism of surface state growth with positive bias is not well understood, but it is possible that it results from a form of solid-state electrolysis, which, as Jorgensen [92] has shown, is capable of reducing SiO_2 to elemental Si, albeit at high temperatures. It would

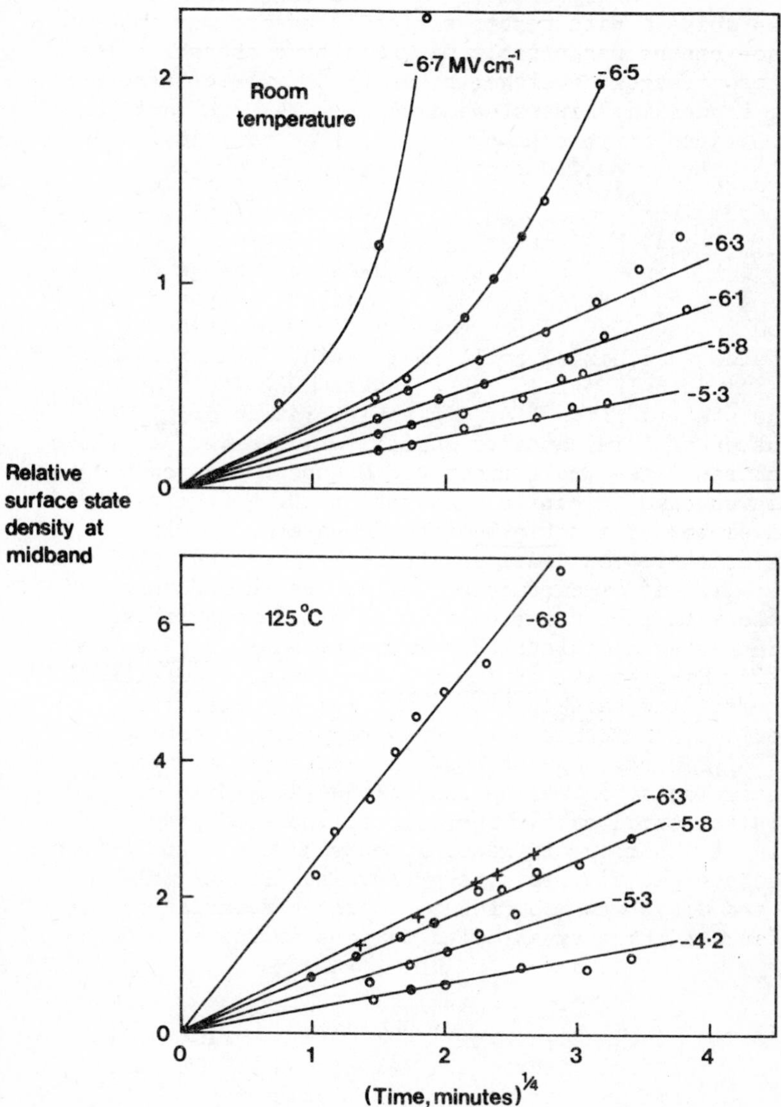

Figure 3.9 Relative increase in the mid-gap density of surface states plotted against $t^{\frac{1}{4}}$, as a function of applied bias and temperature. (After Jeppson and Svensson [90a]).

be extremely interesting to make a study of the interface region after a period of B-T stressing using some of the surface analysis techniques described in Section 1.5.

In conditions where large current densities flow through the $Si\text{-}SiO_2$ interface, the rate of growth of surface states is much faster than the logarithmic time dependence reported earlier. For MNOS devices, which use large gate voltages of either polarity to inject carriers into the insulating layers, Woods and Tuska [94] have found that ΔD_{it} is approximately proportional to the total charge flow through the oxide. This suggests that in these circumstances the states are not caused by trapping in existing defects, but rather by the creation of damage throughout the interface by the high-energy charge carriers.

The growth of surface states in devices exposed to radiation is usually attributable to the production of mobile holes in ionizing collisions, rather than to damage created in the bulk oxide. Thus Hughes [93] found that the final state density (for a given radiation dose) was dependent on the bias applied during irradiation, saturating at a high level for small positive voltages, and reducing as the bias was taken negative. Later work by McGarrity et al [38] showed that for wet oxides (which are much less tolerant of radiation treatment than dry ones) has shown that the build-up of surface states occurs in a much longer time-scale than that required for the holes to drift to the interface. They postulate that when the holes are trapped by breaking Si-H bonds, surface states are not formed until the hydrogen ions have drifted away from the interface. In dry oxides, where the holes are trapped by breaking strained Si-O bonds, the states are created immediately.

Ion implantation, as well as producing radiation damage in the oxide and in the bulk silicon (and thus increasing the number of trapping centres) can also introduce surface states at discrete energy levels. For example, Fahrner and Goetzberger [70] have considered implanted Be, whilst Wang and Evwaraye studied devices implanted with Au.

4

Mobile Ions

4.1 Introduction

When, in the early days of planar processing, the density of surface states had been reduced to a sufficiently low level to allow working MOS transistors to be fabricated, it was found that the threshold voltages were considerably more negative than predicted by simple theory. This led to all p-channel transistors being enhancement types with large (negative) threshold voltages, and all n-channel ones being depletion (or normally-on) type. Moreover, the value of the threshold voltage (for both transistor types) was very unstable, becoming more negative when the gate was biased positively. The instability was observable at room temperature and was strongly thermally activated, but the devices were relatively stable under negative gate bias. These characteristics can be explained by the presence of mobile positive ions in the oxide. Since commercially acceptable devices could not be produced until this instability had been removed, much effort has been expended to identify the mobile species and then either to remove it from the system or to prevent it affecting the device characteristics by immobilising it or rendering it electrically neutral.

4.2 Basic Theory

One of the earliest studies of ion transport in MOS devices was presented by Snow et al [95] in 1965. They considered the generalized case of an arbitrary, one-dimensional charge distribution $\rho(x)$ in the oxide, as is shown in Figure 4.1. The total charge induced in the semiconductor and the metal, Q_s' and Q_g' respectively, are

$$Q_g' = \int_0^d \frac{(x-d)}{d} \cdot \rho(x)\, dx \quad \text{and } Q_s' = \int_0^d \frac{x}{d} \cdot \rho(x)\, dx \tag{4.1}$$

The total mobile charge in the oxide $Q_m(tot)$, is given by

$$Q_m(tot) = \int_0^d \rho(x)\, dx = Q_g' + Q_s' \tag{4.2}$$

Now a gate voltage of $-Q_s'/C_{ox}$ would be required to induce a charge of Q_s' in the semiconductor and hence the effect of the oxide charge distribution is to produce a parallel shift of the C-V curve (with respect to the ideal case) along the voltage axis of

$$\Delta V = \frac{1}{dC_{ox}} \int_0^d x\rho(x)\, dx \tag{4.3}$$

Thus it can be seen that changes in the distribution of the oxide charge during stressing will result in a change in transistor threshold voltage, or capacitor flat-band voltage, even if the total charge remains unchanged. In practice, the distribution $\rho(x)$ is generally unknown, so the effect of the total charge $Q_m(tot)$ is expressed as an equivalent areal charge Q_m at the oxide/semiconductor interface.

An independent measurement of $\Delta Q_s'$ during high temperature biasing can be obtained by integrating the current flowing in the external circuit. If $Q_m(tot)$ is fixed and the only current flowing is ionic (as will be the case for low and moderate electric fields), then

$$\int_0^t i\, dt = C_{ox} V_G + \Delta Q_G' \tag{4.4}$$

where $C_{ox} V_G$ is the capacitor charging (displacement) current, and $\Delta Q_g' = \Delta Q_s'$.

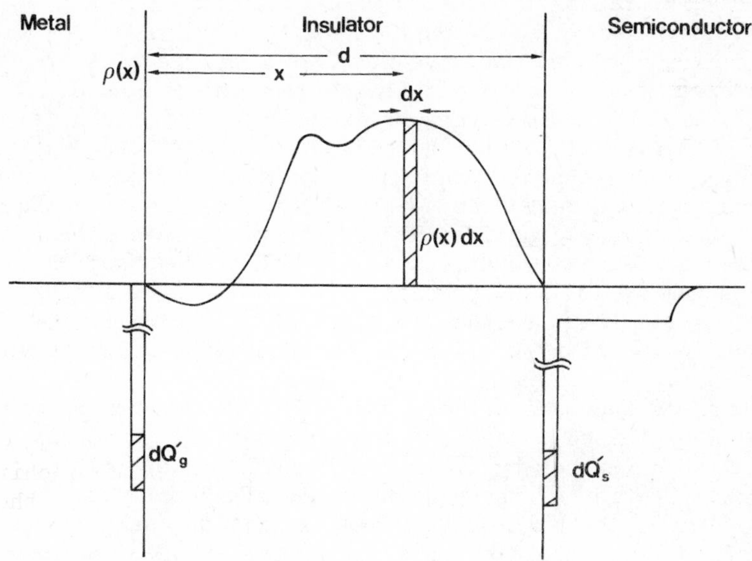

Figure 4.1 Charge induced in the gate metal and in the semiconductor substrate by an arbitrary charge distribution ρ(x) in the insulator.

The number of mobile ions drifting from one electrode to the other revealed by the voltage shift of the C-V curve will only be the same as that found by the integrated current method if the electrodes are <u>blocking</u>, that is, no charge exchange occurs at the electrode, and the mobile species remains in ionic form. It is generally observed [95, 96] that silicon and aluminium (and probably gold as well) form blocking contacts with SiO_2, but in certain cases [97, 98] at least some of the mobile ions exchange charge with the silicon. It should be noted that experiments in which the same quantity of charge is drifted back and forth across the oxide by cycling the bias do not necessarily prove the contacts to be blocking, because the ions could be neutralized and then re-ionized at either interface, depending on the polarity of the applied field.

4.3 Identification of the Mobile Species

Identification of the mobile ions responsible for the drift in MOS threshold voltages rested until fairly recently on indirect evidence, but despite this the terms "mobile ions" and "sodium contamination" have become almost synonomous in the literature. Other mobile ions have been proposed, principally protons (Hofstein [96]) and, more recently, potassium (Guthrie et al, [99]), but Na^+ remains by far the most commonly quoted positive ion, and negative ones are only rarely reported. Many radiochemical investigations of the dynamics of sodium ion transport have been made, such as those by Yon et al [100] and Carlson et al [101] who performed neutron activation and sodium radiotracer experiments, and were able to show that the injection and drift of active sodium ions are superimposed on a stable background of inactive sodium distributed throughout the bulk of the oxide. The presence of quantities of immobile sodium is important because it admits the possibility of another instability should it become activated under stress conditions.

Much of the indirect evidence of sodium contamination [95, 102] is based on showing that MOS devices purposely contaminated (either by immersion in NaCl or NaI solutions, or by evaporation of these salts immediately prior to the deposition of the metal electrode) exhibit the same instabilities as those observed in nominally uncontaminated samples. Direct evidence of sodium ion drift has been produced by an ion microprobe (SIMS) investigation by Guthrie et al [99], who also found potassium in their samples. The concentrations and chemical states of the ions could not, unfortunately, be established with this technique, which can also produce a redistribution of the impurities by the action of the ion beam. The presence of sodium ions in thin oxides not purposely contaminated has also been reported by Grunthaner and Maserjian [103], who employed x-ray photoelectron spectroscopy (XPS) to determine the chemical state of the ions, as will be discussed later.

4.4 Electrical and Thermal Characteristics of Mobile Ion Drif

Figure 4.2 shows the time and temperature characteristics of sodium drift as observed by Snow et al [95], who used a single sample and moved the ions repeatedly from one interface to the other. The effective charge at the Si/SiO_2 interface increases with $t^{\frac{1}{2}}$ at short times, eventually saturating as all the ions are transferred. This time dependence has since been confirmed by many authors [96, 104, 105] and was originally thought to be the result of field-enhanced diffusio through the oxide. However, these later authors have taken the now widely held view that sodium is held in traps situated at both interfaces, and that the ion current is emission-limited, rather than space-charge limited. In the emission-limited model, the $t^{\frac{1}{2}}$ time dependence results from a small Gaussian spread in the trap depth, an assumption also made to

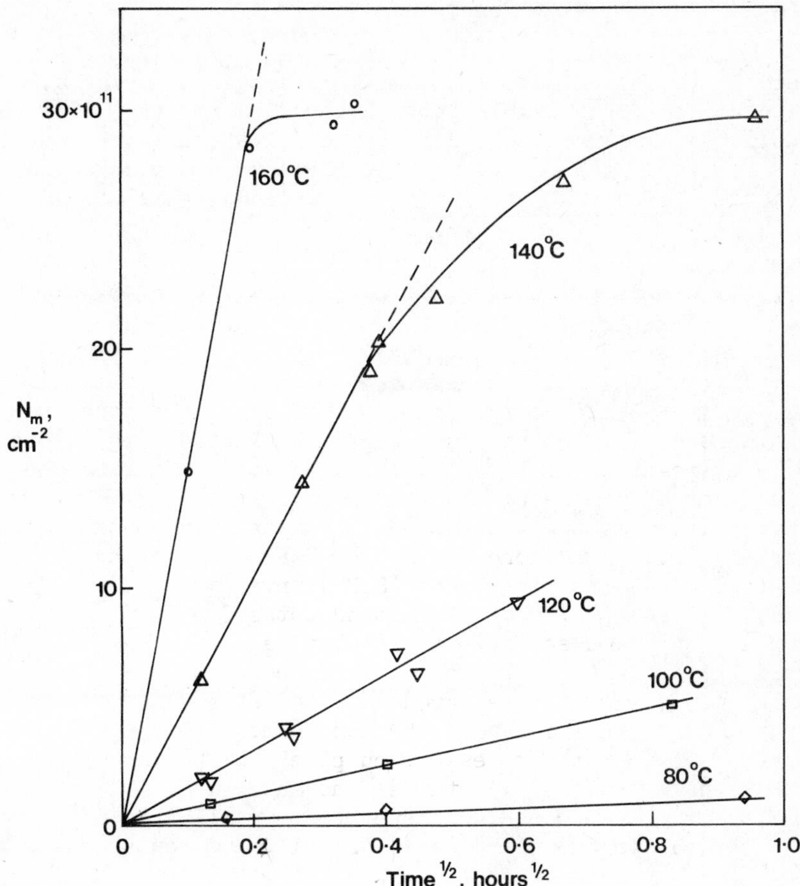

Figure 4.2 Dependence of the excess charge induced in the silicon on the bias time and temperature. Applied field = 5.10^5 V/cm (After Snow et al [95]).

explain the detailed shape of the TSIC curves [59, 106].

The field dependence of the drift is shown in Figure 4.3, and was explained by Eldridge and Kerr [102] by an effective lowering of the trap depth, allowing the combined field and temperature dependence to be described empirically by

$$\Delta Q_s' = K_o Q_o t^{\frac{1}{2}} \exp \frac{-\phi + \frac{1}{2}q\,Ew}{kT} \qquad (4.5)$$

where K_o is a constant, ϕ is the trap depth, E is the electric field and w is the trap width. Later authors [105, 107] have considered that the field dependence is determined by Schottky barrier-lowering, so that the effective trap depth varies with the square root of the field. This model is a good

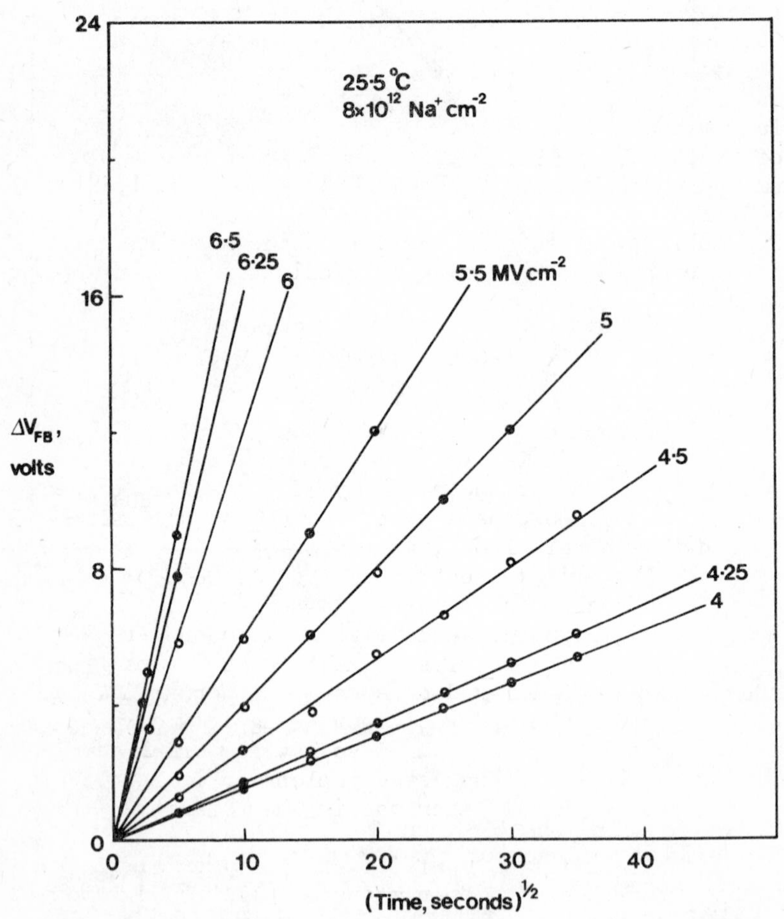

Figure 4.3 Electric field dependence of mobile ion drift at room temperature and with the aluminium biased positively. Sodium ion contamination = 8×10^{12} ions/cm^2. (After Rai and Srivastava [105]).

approximation to the experimentally observed curves at moderate electric fields and sodium concentrations, but at higher concentrations a second, deeper trap appears to become active. However, an alternative explanation is that when the number of ions at the interface is high (as it will be for purposely contaminated samples) the current is space-charge limited (as in Snow's original model), and that only when there are relatively few ions present does the current become emission-limited.

Some confusion exists in the literature over the value of the activation energy of mobile ion drift, as some authors quote values of the trap depth ϕ whilst others (taking a device oriented viewpoint) consider the time required for a given threshold voltage shift and hence (because of the $t^{\frac{1}{2}}$ time dependence) give a value equivalent to 2ϕ. Values of ϕ for sodium drifting from the aluminium to the silicon interface range from 0.49 eV [102] to 0.89 eV [105] with a mean of about 0.7 eV [95]; corresponding mean values for lithium and potassium are 0.48 eV [95] and 0.82 eV [59] respectively.

The mobility of ions in the bulk oxide has been determined by Stagg [108] using a transient ion current technique as

$$\mu(Na^+) = 1.0 \exp(-0.66 \text{ eV}/kT) \text{ cm}^2/\text{V sec}$$

$$\mu(K^+) = 0.03 \exp(-1.09 \text{ eV}/kT) \text{ cm}^2/\text{V sec}$$

Very similar results for the mobility of potassium ions at temperatures above 300°C have been obtained using the TVS method by Hillen et al [108a], who also point out that theoretically the pre-exponential constant should be a function of the reciprocal temperature.

Woods and Williams [109] have injected a number of ionic species into an uncovered oxide by means of a corona discharge, and found that the mobility decreased in the order Li, Na and Cs; that is, in the order of increasing atomic radii. Indeed, the mobility of caesium is so low at normal temperatures that Greenwood [110] has suggested implanting it into field oxides to give a permanent increase in the parasitic threshold voltage.

The lateral diffusion coefficient of sodium along the Si/SiO_2 interface has also been measured [111], and found to be

$$D(Na^+) = 0.003 \exp(-0.8 \text{ eV}/kT) \text{ cm}^2/\text{sec}$$

which is typically much smaller than the values derived from the bulk mobility, and hence supports the view that the ions are held at the interface by traps.

4.5 Evidence for Potassium Motion and Charge Neutralization

Both the thermally stimulated ionic current (TSIC) and the triangular voltage sweep (TVS) techniques often produce two peaks in the oxide current, the first peak being at low temperature or near zero volts, and the second at high temperature or high voltage. Many authors [58a, 59, 106, 108] have ascribed these peaks to sodium and potassium respectively, basing these assignments largely on experiments with purposely contaminated samples. However, Tangena et al [112, 113] and Hino and Yamashita [97] have challenged this viewpoint, and stated that the area under both peaks can be increased by sodium contamination. In the TVS experiments of Tangena, who used a grid electrode structure to allow the contamination to be performed after patterning of the electrodes, the first peak was assigned to a space-charge limited current, and the second to an emission limited current. Hino found that the total area under the TSIC peaks when the applied bias was positive did not correspond to the shift in the C-V characteristics, atlhough the same total charge was observed for repeated sweeps with negative and positive bias. This result implies that some of the mobile ions are neutralized at the Si/SiO_2 interface, but are re-ionized on the subsequent sweep with negative voltage. When the sample was stressed for a period with positive bias, it was found that, when the ions were swept back to the aluminium, the high temperature peak had grown at the expense of the low temperature one, thus suggesting that the high temperature peak (or the high voltage one in TVS experiments) results from the re-ionization of neutralized sodium.

The presence of two forms of sodium at the Si/SiO_2 interface was originally suggested by the results of Hofstein [96], who found a fast and a slow component in the recovery at high temperature of samples previously stressed with positive bias, and is supported by the XPS study of the chemical state of the sodium performed by Grunthaner and Maserjian [103]. They found five distinct peaks in the electron spectrum which could be attributed to various forms of sodium. The first was characteristic of elemental sodium, and showed no response to B-T stressing. The other four peaks were assigned to two states located at each interface, and it was found that one pair grew at the expense of the other during stressing. The pair of peaks at the silicon interface were widely separated (thus accounting for the two recovery mechanisms), and the relative intensities of the peaks varied with stressing as more ions were neutralized or re-ionized.

The form in which the mobile ions are held at the aluminium interface is important because it determines the kinetics of threshold voltage shift in service conditions. Hickmott [57] showed that for samples without a post-metallization anneal the TSIC peak for the first heating with positive bias occurs at higher temperatures than for subsequent sweeps. This is attributed to a chemical reaction between the aluminium and the silicon dioxide in which the oxide is reduced to elemental silicon, and the sodium is moved to shallower traps. In practical manufacturing processes the sintering treatment performed to produce ohmic contacts is likely to produce these relatively shallow traps before the devices reach service. Boudry and Stagg [58a] have shown that there is a range of trap depths at both interfaces, and that when the ions are held at the SiO_2/Al interface by negative bias they can be transferred to deeper traps by increasing the temperature. In this experiment, the curves were reproducible as the ions were swept back and forth, suggesting that although the range of shallow traps are created by high temperature treatment, their occupancy depends on the temperature history of the device.

4.6 Prevention of Mobile Ion Instabilities

4.6.1 Processing conditions

There are three major approaches to solving the problems of mobile ion instabilities; (i) drastically reducing the number of mobile ions present in the oxide, (ii) rendering them immobile, or (iii) electrically neutralizing them.

To prevent alkali ions being incorporated in the oxide basically only requires ultraclean processing techniques, such as the use of ultrapure reagents, careful cleaning of all handling equipment (and minimizing the amount of handling) and well defined washing and preparation techniques. Also, of course, it must be ensured that the silicon substrate itself is not contaminated. Mayo and Evans [114] have studied the impurity content of quartz furnace tubes, and the rate at which sodium diffuses out of them at normal oxidation temperatures. They found that significant amounts of sodium could be leached out, especially when the oxidation ambient was wet oxygen, but it seems possible that some, if not all of this sodium, which is distributed throughout the oxide, is incorporated in an elemental form, and plays no part in the positive bias instability. Nevertheless, it is known that lightly etching the furnace walls with HCl gas prior to oxidation reduces the total sodium content of the oxide, and is thus a useful precaution.

Electrical measurements (which show that the oxides are initially stable under negative B-T stessing), etch-back experiments, and photo I-V measurements [115] all indicate that the mobile ionic charges initially reside at the oxide/metal interface. This points to the metallization step as being one of the most critical processes, since even with the most extreme cleanliness during oxidation, mobile charge densities in the 10^{13}-10^{14} cm^{-2} range are still observed if the metallization process is not equally well controlled. Lightly etching the oxide immediately prior to metallization does not reduce the mobile ion concentration, showing that it is the metallization process itself and not the transfer of the devices from the oxidation furnace to the evaporator that introduces the ions. The biggest single improvement in Q_m occurred when the evaporation of aluminium by resistance heated tungsten filaments was replaced by electron beam heating. This change made it possible to reduce N_m to below 10^{10} cm^{-2} (albeit at the expense of a number of technological problems, such as radiation damage and rather small grained aluminium films) so that devices can now be made with threshold voltage shifts due to mobile ion drift of as little as 10 mV.

4.6.2 Phosphosilicate glass

As early as 1964 [116] it was found that phosphosilicate glass films (PSG) formed on a thermally grown SiO_2 layer produced a marked increase in the surface stability of silicon planar devices. The glass has the property of gettering mobile ions from the oxide and trapping them, greatly reducing their subsequent drift under positive B-T treatment. Also, the glass acts as a barrier to the introduction of alkali ions during later processing steps, and during service. The deposition and stabilizing properties of the glass have been reported by Eldridge and Kerr [102]. Unfortunately, PSG is polarizable [117] (as described in Chapter 5) and so great care needs to be taken in determining the optimum thickness and composition of the glass. The phosphosilicate glass is usually formed by adding P_2O_5 to the thermal oxide by pyrolysis of P_2OCL_3 or PH_3 [118] in O_2, followed by heat treatment at high temperature (~1000°C) in nitrogen. Generally P_2O_5 concentrations in the range 1-10 mole per cent are used with a thickness of approximately 10% of the total insulator thickness. Sodium ions introduced onto the surface of the glass, before metallization, drift through it in much the same way as through pure oxide, but at a much slower rate. Eldridge and Kerr [102] have reported a $(time)^{\frac{1}{2}}$ dependence, and have also shown that the activation energy is dependent on both the initial sodium concentration and the glass composition, as is shown in Figure 4.4. The reason for these

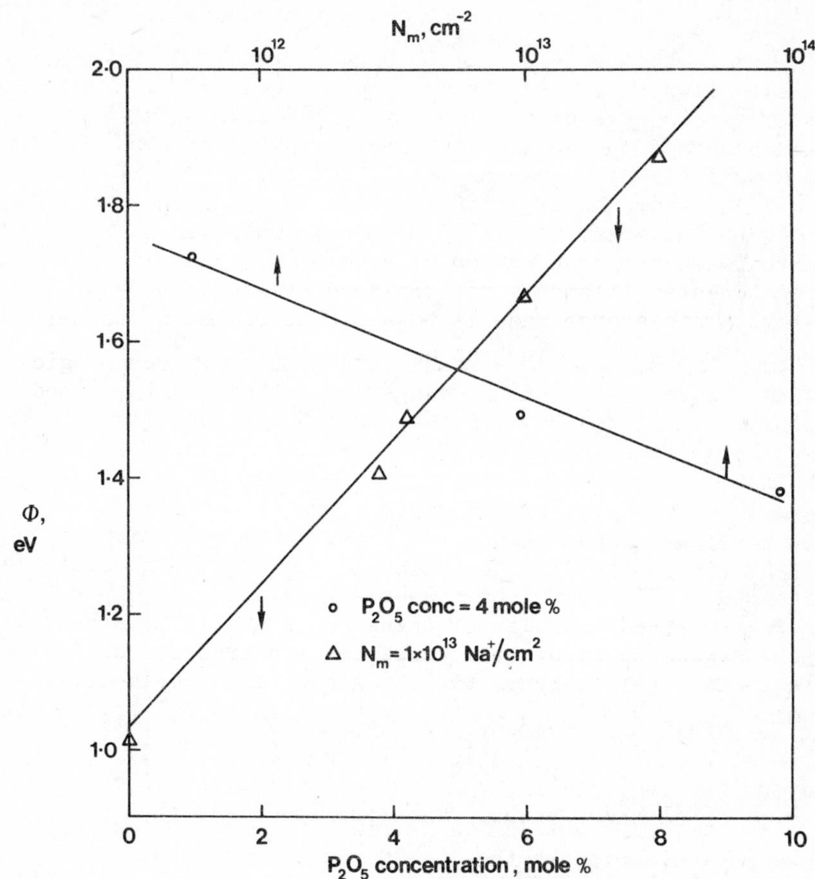

Figure 4.4 Dependence of activation energy ϕ on PSG composition for a constant sodium concentration and vice versa. (After Eldridge and Kerr [102].)

dependencies is that the sodium ions are trapped in potential wells whose depth and width are a function of the average distance between sodium ions, and of the P_2O_5 concentration. A structural model for the traps is based on the replacement of the SiO_4 tetrahedra found in the somewhat randomized network of a pure oxide with tetrahedral PO_4 groups and a non-bridging negative oxygen ion associated with each pair of tetrahedra. The two tetrahedra are found in close association with each other (as evidenced by the polarizability of the bulk glass), so that the sodium is trapped by coulombic attraction, as shown schematically below in Figure 4.5.

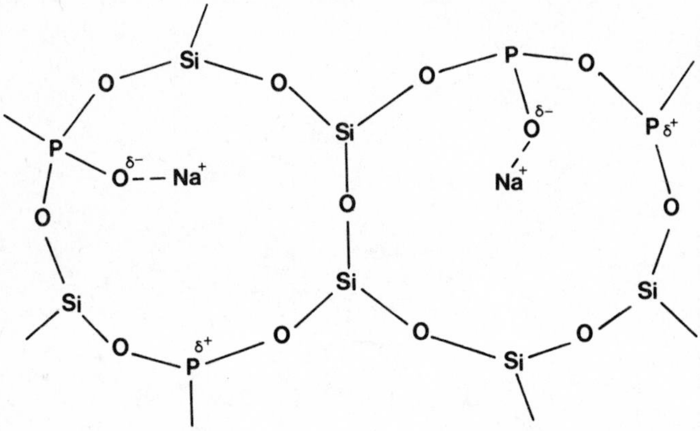

Figure 4.5 Schematic two-dimensional model of the phosphosilicate glass structure, showing the mechanism of the Na^+ trap via coulombic interaction.

The practical results of using PSG stabilization have been predicted by considering the time required for a -0.1V shift in threshold voltage at $80^{\circ}C$ and under a field of $+2 \times 10^6$ V/cm, and are shown in Figure 4.6 as a function of the total sodium concentration. It must be pointed out, however, that the stability of thermal oxides has been improved since these results were published, so that a sodium contamination in the 10^9-10^{10} cm^{-2} range, as is currently achievable, does not in itself produce unacceptable threshold voltage shifts.

4.6.3 Chlorinated oxides

If mobile ions cannot be completely irradicated from the MOS system, then the next best solution is to neutralize them by producing non-blocking electrodes. One possible means of achieving this is to use wet oxides, which have a considerable numbers of non-bridging oxygens at the interface; unfortunately however, steam-grown oxides do not always show the predicted stability for two practical reasons. Firstly, the steam itself may be contaminated, and secondly it can leach sodium from the furnace walls.

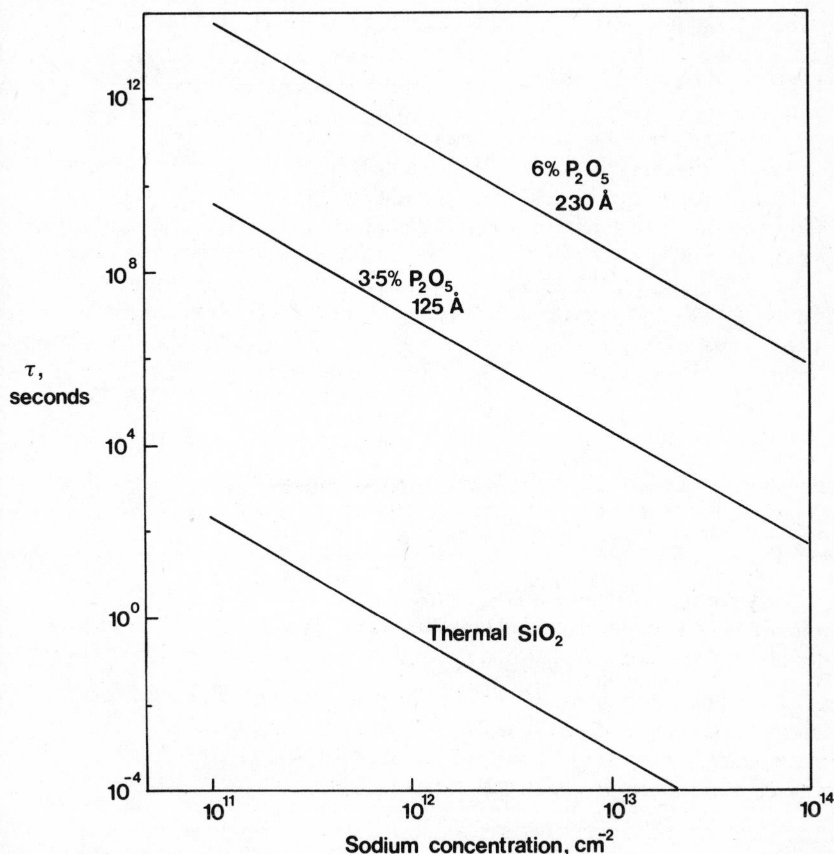

Figure 4.6 Dependence of the estimated time, τ, required for a sodium induced threshold voltage shift of -100 mV, for various PSG passivating layers calculated for a gate thickness of 1000Å, at 80°C and with a field of +2 x 10^6 V/cm. (After Eldridge and Kerr [102]).

The use of chlorinated oxides as a means of reducing the mobile ion concentration was discovered in 1972 by Kriegler et al [119], who found that samples contaminated with NaCl had fewer mobile ions than those contaminated with Na_2CO_3.

Chlorinated oxides can be produced by introducing small amounts (1-10%) of HCl or Cl_2 into the oxidizing ambient, or by bubbling the oxygen through trichloroethylene, but the first method is by far the most widely used. The structure of HCl oxides has been studied by Monkowski et al [120] who found chlorine rich islands with a diameter of the order of 1 µm at the Si/SiO_2 interface. With large HCl concentrations (or with Cl_2), and after long periods at the oxidation temperature, these islands develop into pockets of gas (possibly a chlorosiloxane) which can lift the oxide and etch the substrate.

The chlorine is incorporated in the oxide in an electrically neutral form, and the mobile sodium ions must be drifted to the Si/SiO_2 interface by a period of positive B-T stress before they are neutralized. The kinetics of the neutralization process have been studied by Rohatgi et al [121] and Stagg and Boudry [98, 111]. The temperature dependence of the neutralization is shown in Figure 4.7 for two different chlorine concentrations; in general, the amount

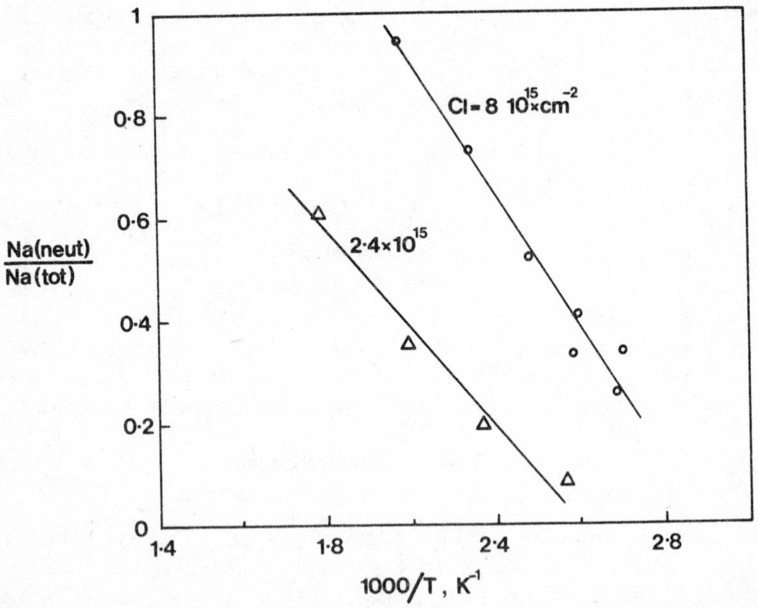

Figure 4.7 Neutralization of mobile sodium against reciprocal temperature, for a 0.5 minute positive B-T stress. (After Stagg and Boudry [98]).

of sodium which can be neutralized is about two orders of magnitude smaller than the chlorine concentration. The time dependence of the neutralization cannot be described by a single time-constant, and is largely independent of the electric field. It has been suggested [111] that the limiting factor for the neutralization is the lateral diffusion of the sodium ions to the chlorine-rich islands.

The major problem with HCl oxides is that the sodium ions are not permeanantly trapped, but can be re-ionized by negative B-T treatment. Also, this method of reducing the mobile charge density requires a "burn-in" period before it is effective, and it still leaves the sodium in the oxide, which may possibly contribute to a later devitrification of the oxide [24]. The preferred use of HCl is thus to pre-clean the furnace tube, and to rely on clean processing to prevent any contamination after oxidation.

Attempts have been made [122] to replace chlorinated oxides with fluorine ions implanted into conventional oxides. Some reduction in the mobile charge density was noted, but this is attributed to gettering by the implant damage, rather than by the chemical action of the ion itself.

4.6.4 Silicon nitride

Silicon nitride (Si_3N_4) is almost completely impervious to ion motion, and is sometimes used as a barrier layer on top of the oxide. It also has the added advantage of a higher dielectric constant than SiO_2 ($\varepsilon_{ox} = 3.9$, $\varepsilon_{nitride} \sim 6.9$) and so increases the gain of MOS transistors. The nitride, which is usually produced by the thermal decomposition of silane and ammonia, does not have the ability to getter contaminants from the underlying oxide (unlike PSG), but neither does it suffer from dipolar polarization. The slow movement of ionic contamination (not necessarily metallic) through a nitride layer has been observed [104] to have a $t^{\frac{1}{2}}$ time dependence similar to that seen in SiO_2, but the MNOS structure also suffers from a "double-dielectric" instability (described in Chapter 8) and so is generally only used in specialized memory transistors.

4.7 Non-Uniform Lateral Impurity Distributions

So far, it has been assumed that the ionic impurity concentration is uniform across the area of an oxide, and thus under B-T treatment the mobile ions drift as a uniform sheet of positive charge, producing a parallel shift of the C-V curve to more negative voltages. Experimentally, however, it is often found that the shift is accompanied by a reduction in slope of the curve which cannot be accounted for fully by a growth in the surface state density. In these cases, the spreading of the C-V curve can be considered as being due to a laterally non-uniform impurity distribution. Silversmith [123] has used a gamma distribution to describe the probability of finding a given ionic density in a small area, and found that the standard deviation of the local density is about 50% of the mean value.

Local variations in the ionic density effect the Si/SiO_2 interface barrier height, a property which DiStefano [124] and Williams and Woods [125] have exploited to map the density by means of the internal photoemission currents. They found small patches with very high densities, a feature which has been predicted theoretically [126] on the basis of a reduction in the electrostatic image potential produced when the ion concentration is high enough to shorten the silicon Debye length.

The clustering of the mobile ions present in an oxide does not affect the threshold voltage directly, but it does have several serious reliability implications. In particular, it is thought [107] to be one of the major causes of "dielectric wearout" (that is, a degradation of the dielectric breakdown voltage, resulting in a catastrophic failure of the oxide) through one of two mechanisms; (i) current runaway resulting from the reduced barrier height and (ii) local devitrification. The much higher than normal oxide current could also cause problems in memory devices in which the oxide is required to isolate the charge stored on a floating node.

4.8 Organic Contamination

Although the vast majority of the literature on ionic contamination deals with metallic ions, failures in integrated circuits sometimes occur by the drift of much more mobile species. These instabilities are not at all well characterized and vary considerably from one process to another, being particularly dependent on the washing and cleaning steps.

In the early studies of ion instabilities, Hofstein [96] considered protons (H^+) to be a major contaminant, but they have since largely been ignored, although Tangena et al [112] have reported increases in both TVS peaks as a result of contamination in acetic acid, and Nemeth-Sallay et al [127] found TSIC peaks at room temperature and below for ether-treated oxides. Nakayama et al [128-130] have observed both threshold voltage and junction leakage instabilities in p-channel silicon-gate transistors stressed for a few minutes with positive gate voltages at room temperature, and have associated these instabilities with washing in ethyl alcohol or acetone. It appears that protons are formed by the chemical breakdown of traces of alcohol or acetone in the source and drain regions during the post-metallization annealing stage (catalyzed by the presence of Al_2O_3 at the Al/SiO_2 interface) and drift into the active device region. The role of the boron-doped polysilicon gate is not understood, but is essential to the instability - no similar instabilities have been reported for phosphorous-doped n-channel transistors, possibly due to the gettering action of the thin layer of phosphosilicate glass formed under the gate. Positive bias stress is an unusual condition to apply to p-channel transistors, but instabilities have also been seen [131, 132] in a wide range of p-channel silicon-gate transistors from commercial processes stressed with negative bias. In these cases, which are also explained by the lateral diffusion of a mobile positive species from the source and drain overlap regions into the active channel, the instability only became apparent after extended stressing at high temperatures, and had an activation energy of 0.7 eV [131], compared to the value of 0.54 eV reported by Nakayama [129].

Other examples of mobile ion induced instabilities tend to be of a "one-off" nature, and are not often reported in the literature. A typical case is the long term reduction in the field threshold voltage of a p-channel MNOS process due to negative ions (which are only very rarely observed) introduced by corrosion in the furnace pipe-work. The ion involved was not identified, but other workers [132a] have found similar a effect, with an activation energy of 1.4 eV, in field oxides grown at high temperatures in wet oxygen. In this case a $t^{\frac{1}{2}}$ time dependence was found, and the instability was removed by phosphorous or HCl gettering. This suggests that the instability might be caused by electron injection and trapping from the aluminium gate electrode, aided by the barrier lowering produced by mobile sodium ions drifted to the Al/SiO_2 interface.

D

5

Dipolar Polarization

5.1 Introduction

The use of phosphosilicate glass (PSG) layers (with a composition SiO_2-P_2O_5) deposited on top of the gate oxide to getter mobile sodium ions [133] and also to present a barrier to their diffusion from external sources was described in the previous chapter. Unfortunately, the molecular structure of PSGs is such that they contain a number of permanent dipoles which can be aligned by an applied electric field, and hence creates another instability mechanism. This instability, which is called "dipolar polarization" here, should not be confused with the so-called "polarization" of all gate insulators, a term which was used loosely in many early papers to include all instabilities.

The main characteristics of dipolar polarization have been investigated by Snow and Deal [117] using dot capacitors as shown in Figure 5.1, and they found that it causes a parallel shift of the C-V characteristics along the voltage axis, with ΔV_{FB} and V_G having opposite polarities (as in the case of ionic drift). Unlike ionic drift, however, it is completely symmetrical and reversable for either polarity of applied bias. The increase of ΔV_{FB} with time is initially approximately logarithmic, but it eventually saturates at a value $\Delta V_{FB}(sat)$ which is a function of the bias voltage (in contrast with mobile ion drift, in which the time constant but not the saturation value is voltage dependent).

Snow and Deal [117] have shown by means of an etch-off experiment that dipolar polarization is confined to the glass layer, with pure SiO_2 being immune to this particular instability. If the polarizability of the glass is χ_p, the

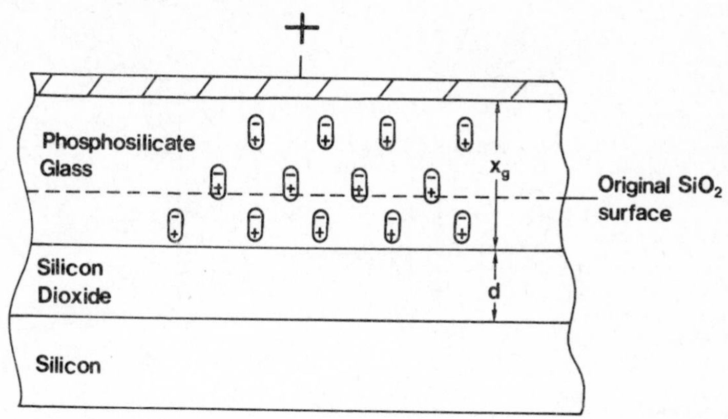

Figure 5.1 Cross-section of a metal-glass-SiO_2-silicon (MGOS) structure, showing dipolar polarization.

volume polarization P eventually reached will be

$$P = \varepsilon_o \chi_p E_g \tag{5.1}$$

where ε_o is the permittivity of free space and E_g is the electric field in the glass. This volume polarization is equivalent to a surface charge density Q_p at the oxide/glass interface, and an equal and opposite charge density on the metal/glass interface. The bias voltage V_G applied to the gate electrode must equal the sum of the voltage drops across the oxide and the glass, ie

$$-V_G = E_g x_g + E_{ox} d \tag{5.2}$$

where E_{ox} is the field in the oxide, d is the oxide thickness and x_g is the glass thickness. Since the electric displacement must be continuous across the oxide/glass interface

$$\varepsilon_{ox} E_{ox} = (\varepsilon_g^u + \chi_p) E_g \tag{5.3}$$

where ε_g^u is the unrelaxed (i.e. high frequency) value of the glass permittivity.

Thus $$Q_p = \frac{-\varepsilon_{ox} \varepsilon_o \chi_p V_G}{(\varepsilon_g^u + \chi_p)d + x_g \varepsilon_{ox}} \tag{5.4}$$

The charge induced in the silicon (Q_s') will be

$$Q_s' = \frac{C_{ox} Q_p}{C_{ox} + C_g} \qquad (5.5)$$

where C_{ox} and C_g are the oxide and glass capacitances respectively.

The total shift in the C-V characteristics is thus

$$\Delta V_{FB}(sat) = \frac{-\varepsilon_{ox} x_g \chi_p}{\varepsilon_g^u (\varepsilon_g^u + \chi_p)d + \varepsilon_{ox} x_g} .V_G \qquad (5.6)$$

For practical PSG films, the glass thickness is often small compared to the total insulator thickness, and the concentration of dipoles (and hence χ_p) is kept low, so Equation (5.6) may be simplified to

$$\Delta V_{FB}(sat) = \frac{-\chi_p x_g}{\varepsilon_g d} .V_G \qquad (5.7)$$

Both Equation (5.6) and (5.7) agree with the finding that the final shift of the C-V characteristics is a fraction of the applied bias.

5.2 Time, Temperature and Processing Dependencies

The dependence of the saturated flatband shift (or, of course, the threshold voltage shift of a transistor) on the relative thickness of the glass is shown by the results of Eldridge et al [134] in Figure 5.2 (where the lines apply for either polarity of the stressing voltage), confirming the validity of Equation 5.6. Figure 5.2 also shows that the polarization effect can be minimized by reducing the concentration of P_2O_5 in the glass although, as shown in Chapter 4, this also reduces its ability to prevent mobile ion instabilities. If the concentration M_p is defined as the mole% of P_2O_5 in the glass, then Figure 5.3 shows that

$$\chi_p \propto M_p^2 \qquad (5.8)$$

for a wide range of concentrations.

The approach of ΔV_{FB} to its saturated value cannot be described by a simple exponential time dependence, but instead requires a range of time constants to describe it fully. Nevertheless, when describing the temperature dependence of

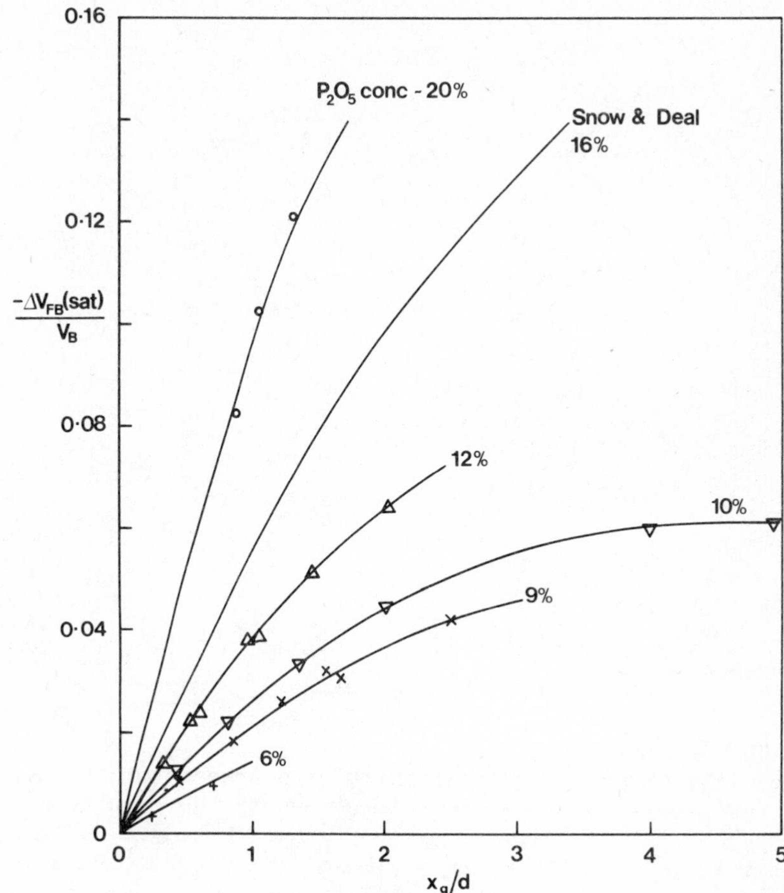

Figure 5.2 Dependence of ΔV_{FB}(sat) per unit of applied voltage on PSG composition and relative thickness. (After Eldridge et al [134]).

the shifts, it is customary to define τ as the time required to reach 1/e times the saturated value, starting from the relaxed state. Two methods exist for measuring τ. The first involves the normal C-V plotting method, and is usually used at fairly low temperatures, whilst the second method, used by Eldridge et al [134], allows measurements to be made at higher temperatures (where τ is in the range 10^{-3}-10^{-6} seconds) by measuring the dielectric loss ($\varepsilon''/\varepsilon'$) of the glass. This

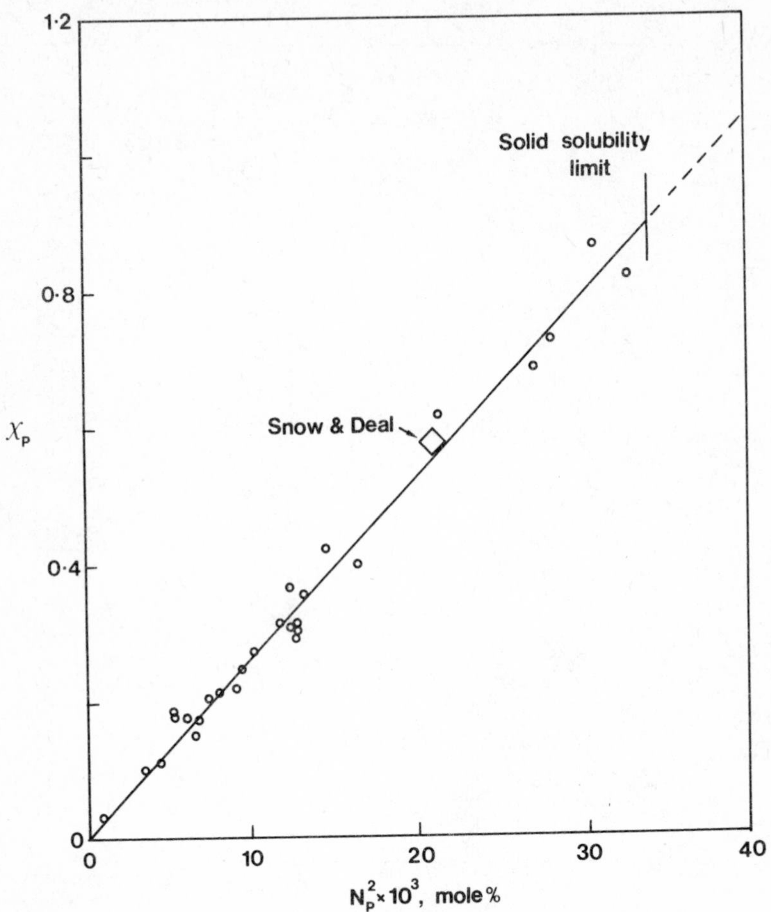

Figure 5.3 Dependence of PSG polarizability on P_2O_5 content. (After Eldridge et al [134]).

loss is measured as a function of temperature and frequency of the applied a.c. signal. If the relaxation were a simple one involving a single time constant, then the loss peak would take the form of the Debye function:

$$\tan \delta = \varepsilon''/\varepsilon' = \Delta\varepsilon(\omega\tau/(1 + \omega^2\tau^2)) \tag{5.9}$$

where $\Delta\varepsilon$ is the relaxation strength ($= \chi_p/\varepsilon_g^u$) and δ is the loss angle by which the polarization lags behind the electric field. The effect of a distribution of τ values results in a broadening of the loss peak, but it will still occur at $\omega\tau = 1$.

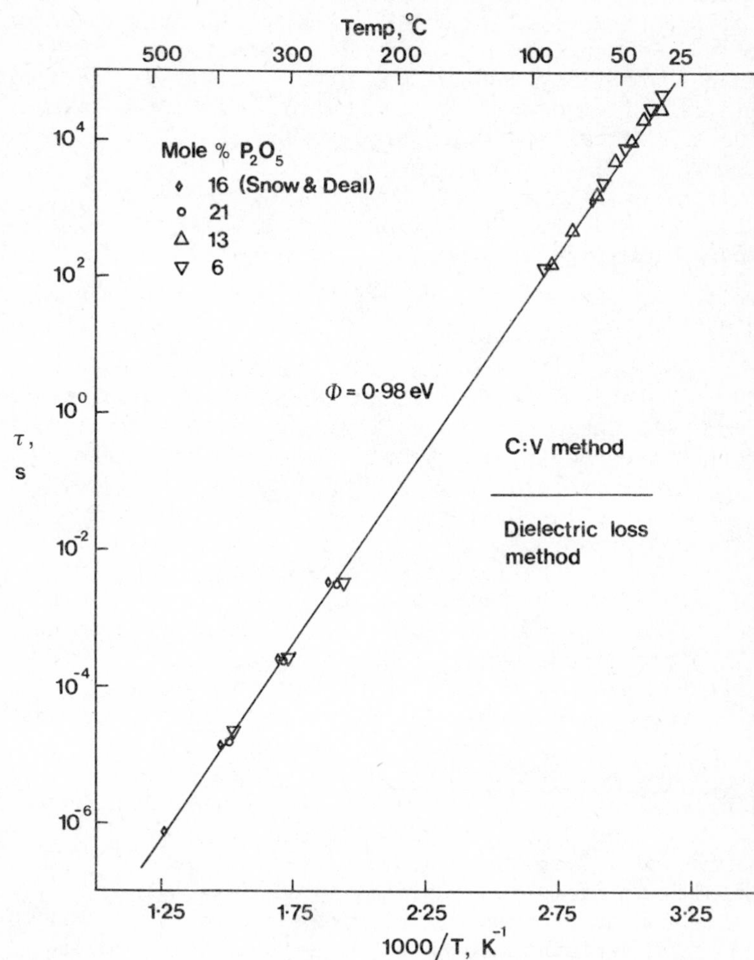

Figure 5.4 Relaxation time as a function of reciprocal temperature. (After Eldridge et al [134]).

By use of the two techniques, the variation of τ over a range of 10 orders of magnitude may be plotted, as can be seen in Figure 5.4. There is excellent agreement between the techniques, and it is also apparent that the glass composition does not effect the time constant of the polarization, a single activation energy of 0.98 eV describing the process over the temperature range 25-500°C. However,

Reynolds [135] has reported a lower activation energy (in the 0.8 eV region) and also a shorter time constant (3 instead of 20 minutes [117] at 50°C) for commercial MOS transistors with a 13 mole % PSG layer approximately 400Å thick. Strictly speaking, the results of Reynolds apply to depolarization, but the process is generally believed to be symmetrical. Of all the common instability mechanisms, with the possible exception of mobile ions from organic contamination, dipolar polarization is by far the fastest to reach saturation under normal operating conditions.

The polarizability χ_p of the glass is also slightly temperature dependent, although practically this is of secondary importance. Eldridge et al [134] have reported χ_p to have a negative temperature coefficient in the temperature range 160-500°C, with glasses with low P_2O_5 content showing the largest dependencies. Reynolds [135] however, observed a positive temperature coefficient in the more practical region from 20 to 80°C but his evidence indicates that the temperature coefficient is reducing with temperature, so that these results are not necessarily incompatible with those of Eldridge.

5.3 "Slow Polarization"

In addition to the fast dipolar polarization already discussed, many authors [117, 118, 134] have reported the presence of a much slower instability in phosphorous-doped SiO_2 films, and discussed it in terms of polarization of the PSG layer. An example of this instability is shown in Figure 5.5 for a thin (60Å) glass layer with only 1.5 mole % of P_2O_5. Both the time constant of the shift and its saturated value are strong functions of temperature; Snow and Deal [117] have quoted an activation energy of 1.15 eV, but Eldridge [134] and Chaudhari [118] have pointed out that the temperature and voltage dependencies are interdependent.

The slow polarization instability is sometimes equated with mobile ion drift through the glass. This is unlikely to be the true explanation, however, because the mechanism is (like the true, dipolar polarization) symmetrical for both polarities of applied voltage, and any mobile sodium ions are initially located at the glass/metal electrode. Although rarely discussed in the literature, a much more probable explanation is that it is a form of the "double-dielectric" instability often reported for MNOS devices (see Chapter 8). The conductivity of PSG is much larger than that of pure SiO_2, and the dielectric constants are very similar [117], so when a bias voltage is first applied to an MGOS device the current through the PSG will be larger than that through the oxide. Thus charge will build up at the glass/oxide interface

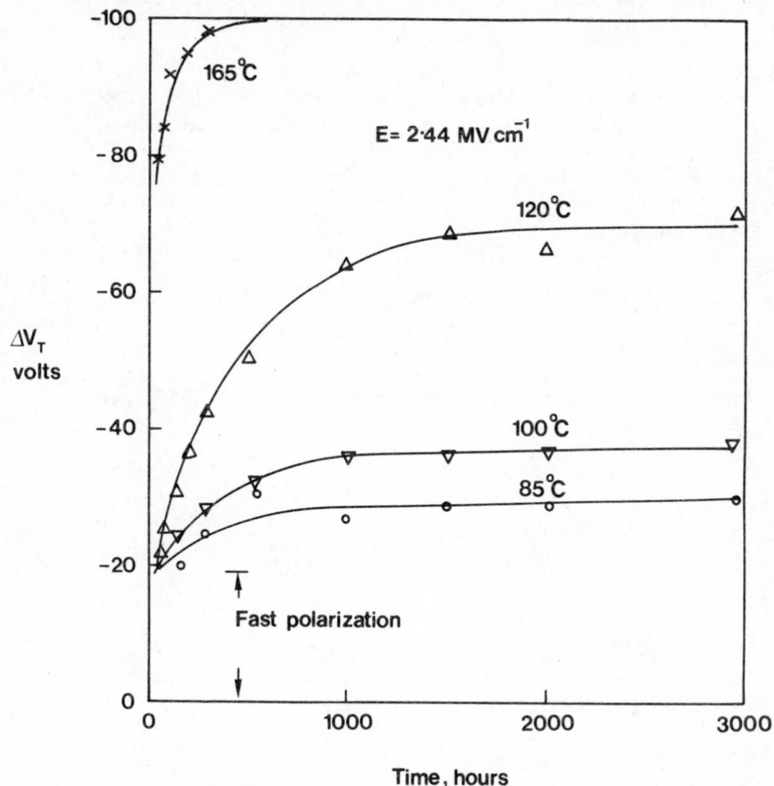

Figure 5.5 Threshold voltage shift as a function of time for phosphorous-doped SiO_2 transistors, due to the "slow polarization" mechanism. (After Chaudhari et al [118]).

to reduce the field in the glass until the currents are equal and equilibrium is reached. An alternative explanation of the slow polarization, proposed by Eldridge [134] is the migration of negatively charged non-bridging oxygen ions over many atomic distances, but it is difficult to reconcile this mechanism with the observed temperature and voltage dependencies of the saturated shifts.

5.4 Non-PSG Polarization

It is difficult to establish whether insulators other than PSG suffer from dipolar polarization because they are invariably used in conjunction with a layer of SiO_2, and have any polarization is obscured by the double-dielectric instability. This is particularly true for silicon nitride, which is generally thought to be free of permenant dipoles as long as it is stoichiometric.

One insulator which is reported [136-138] as showing polarization is Al_2O_3, which can be vapour deposited to provide a barrier to sodium ions, and which also produces a fixed negative charge to offset the fixed positive charge at the Si/SiO_2 interface. The response of a 500Å Al_2O_3-1000Å SiO_2 structure to a one hour stress at various temperatures and voltages (sufficient to cause saturation of the flat-band shift) is shown in Figure 5.6, where it can be seen that the devices also exhibit mobile ion drift with positive bias,

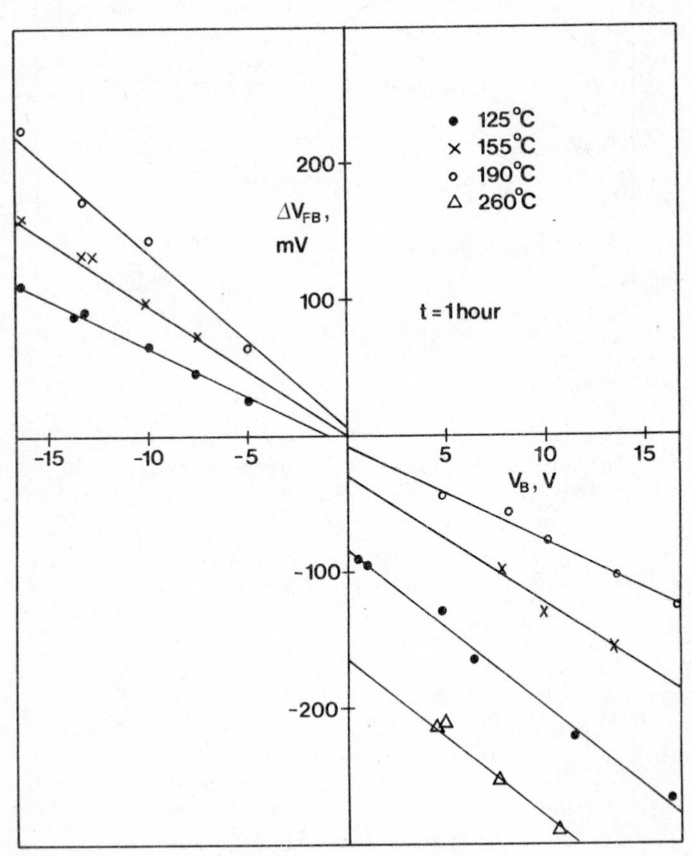

Figure 5.6 Flat-band voltage shift (ΔV_{FB}) for Al_2O_3-SiO_2 dielectric after 1 hour B-T stress (After Gnadinger and Rosenzweig [138]).

possibly due to a species trapped at the SiO_2/Al_2O_3 interface. Lampi and Labuda [137] studied a similar structure and found that the value of ΔV_{FB}(sat) had an activation energy of ~0.3 eV, a value which is higher than expected and leads to the possibility that the shifts are at least partly due to charge trapping (despite the view [136] that electron injection into Al_2O_3 is negligible at the low fields involved).

5.5 Origins of Dipoles in PSG

Although Snow and Deal originally thought that the fast instability in PSG films was due to the presence of a small amount of a conductive phase (micro-crystallites) randomly distributed throughout the glass, no direct evidence has ever been found for the existence of these regions, and the generally accepted model for polarization is that due to Eldridge et al [134]. In this model the glass is formed by the replacement of SiO_4 tetrahedra in the pure oxide with PO_4 groups, which exist in pairs with one negatively charged, non-bridging oxygen ion for each P_2O_5 molecule. Thus, as is shown schematically in Figure 5.7, when an electric field is applied across the glass the oxygen ions can jump from one phosphorous group to an adjacent one, producing a redistribution

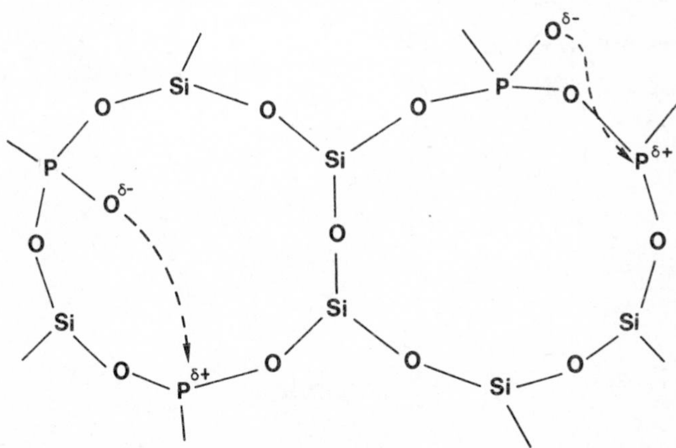

Figure 5.7 Two-dimensional model for the PSG structure, showing the mechanism for dipolar polarization. (After Eldridge et al [134]).

of charge throughout the glass. This mechanism is consistent with the observed quadratic dependence of χ_p on the phosphorous concentration because, for relatively dilute solutions, the probability of having one of each of the two types of PO_4

tetrahedra close enough for the oxygen ion to jump between them is proportional to M_p^2. The distribution of time constants is explained by local variations in the configuration of the dipoles, such as non-uniformity of the bond angles and lengths.

The migration of oxygen ions was also used in Section 5.3 as a possible explanation of the "slow polarization" mechanism. In this model all of the oxygen ions would eventually accumulate at one of the glass interfaces, so that the saturated flat-band shift would depend on the total number of P_2O_5 molecules present.

6

Hole Trapping

6.1 Introduction

Historically, if it is said that the presence of surface states prevented early workers fabricating working MOS transistors, and mobile ion instabilities later delayed the introduction of commercial devices, then the instability now widely described as "slow hole trapping" is mainly responsible for the continued difficulties in producing high-voltage p-channel devices with high reliability.

The principal properties of this instability were succinctly described by Nicollian [139] in 1974. With negative bias applied to the gate electrode, the flat-band voltage drifts steadily in a negative direction, but with positive bias there is a negligibly small positive shift. This behaviour is not only a reversal of the stress-voltage asymmetry observed for mobile ion instabilities, but also the sense of the shift is reversed ($\Delta V_{FB}/V_G < 0$ for ion drift and dipolar polarization, $\Delta V_{FB}/V_G >$ for hole trapping). Thus the hole trapping mechanism, which is often referred to as the negative-bias instability, is characteristic of charge injection into the oxide, rather than a rearrangement of charge. In many aspects of its behaviour, hole trapping may be regarded simply as an increase in the fixed (positive) oxide charge Q_f with negative B-T stressing. In practice, it is usually accompanied by an increase in the density of surface states [140] so the increase in the absolute value of the threshold voltage shift for a p-channel transistor is larger than the flat-band shift, whereas for the corresponding n-channel transistor it is smaller because the two effects work in opposition.

6.2 Time, Temperature and Voltage Dependencies

One of the earliest (1967) studies of the time, temperature and voltage dependence of slow hole trapping is that of Hofstein [141], who found that the threshold voltage initially increased linearly with logarithmic time, but eventually saturated at a value which was almost independent of the temperature. The activation energy of the process was 1.0 eV. The voltage dependence of the time constant of the shift was relatively small (compared to ion drift, for example), but the saturated value of ΔV_T was linearly proportional to the electric field, at least for low to moderate values. Steam grown oxides on (111) silicon were the most susceptable to this mechanism, and dry oxides on (100) the least.

Since Hofstein's early paper, many investigators [142-145] have reported similar findings. Indeed, it can be said that all p-channel technologies (for which negative bias is the normal operating condition) suffer from hole trapping to some extent, although modern fabrication conditions have made significant reductions in the effect, so that, even with accelerated ageing, saturation is rarely observed. For example, some results obtained by Reynolds [145] on a commercial integrated circuit produced by a fairly primitive technology are shown in Figure 6.1. In the absence of a detailed model for the trapping mechanism, the following empirical expressions were fitted to the observed shifts:

$$\Delta V_T = \Delta V_{sat} \left[1 - \exp\left[-(t/\tau)^n\right]\right] \quad (6.1)$$

where

$$\Delta V_{sat} = \Delta V_o \exp(-q\,\theta/kT) \quad (6.2)$$

and

$$\tau = \tau_o \exp(q\phi/kT). \quad (6.3)$$

The variables n, θ, ϕ, ΔV_o and τ_o were adjusted to fit the data. Values of 1.0 eV and 0.09 eV for ϕ and θ respectively are in agreement with Hofstein's results, and are generally accepted [146] as being characteristic of hole trapping, although Sinha and Smith [140] found the activation energy of the shift at short times ($<$ 16 hours) to be only 0.64 eV.

There is considerable disagreement on the exact form of the time dependence of the shifts. For stressing periods much shorter than the time to saturation, Equation 6.1 predicts a t^n time dependence. Reynolds [145] found n to be 0.33, and Sinha and Smith [146] used a value of 0.2. These values are in

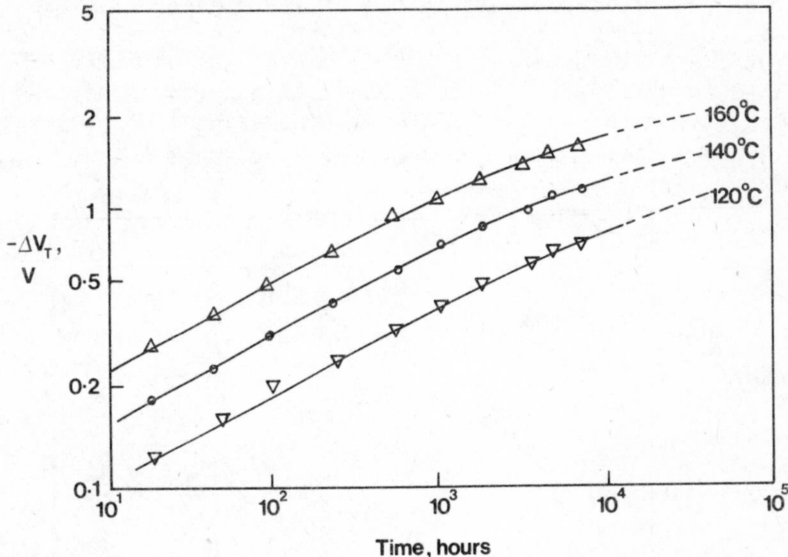

Figure 6.1 Increase in the magnitude of the threshold voltage of a p-channel MOST with negative B-T stressing. (After Reynolds [145]).

agreement with the theoretical model of Jeppson and Svensson [90a], which was introduced in Chapter 3 to account for the growth of surface states, and which gives rise to a $t^{\frac{1}{4}}$ time dependence at moderate fields. However, other workers [146] have observed values of n in the range 0.2 to 1, depending of the processing technology used. Based on Walden's [148] theory of charge trapping in an insulator, many authors [143, 144, 149] have plotted the threshold voltage shift on a linear scale against logarithmic time, and obtained reasonable straight lines. However, as was pointed out by Jeppson, for reasonably stable devices there is very little difference between semi-logarithmic behaviour and a power law with a fractional power, at least over a few decades of time.

Although Hofstein reported a linear dependence of the rate of threshold voltage shift on the applied voltage, later authors have found a much stronger dependence, especially near room temperature. Jepsson and Svensson [90a] found an exponential relationship, whilst Sinha and Smith [140], whose results are shown in Figure 6.2, observed another power law dependence with the power m varying from 1.5 at 300°C to

3 at 50°C. Both of these two relationships give a reasonable fit to the experimental data at low and moderate applied bias, but when the electric field in the oxide exceeds 6–7 MV/cm the voltage dependence becomes much stronger. This field corresponds to a change of the time dependence to a linear one, and also to the onset of strong conduction through the oxide, and thus seems to indicate that a second mechanism is operative here.

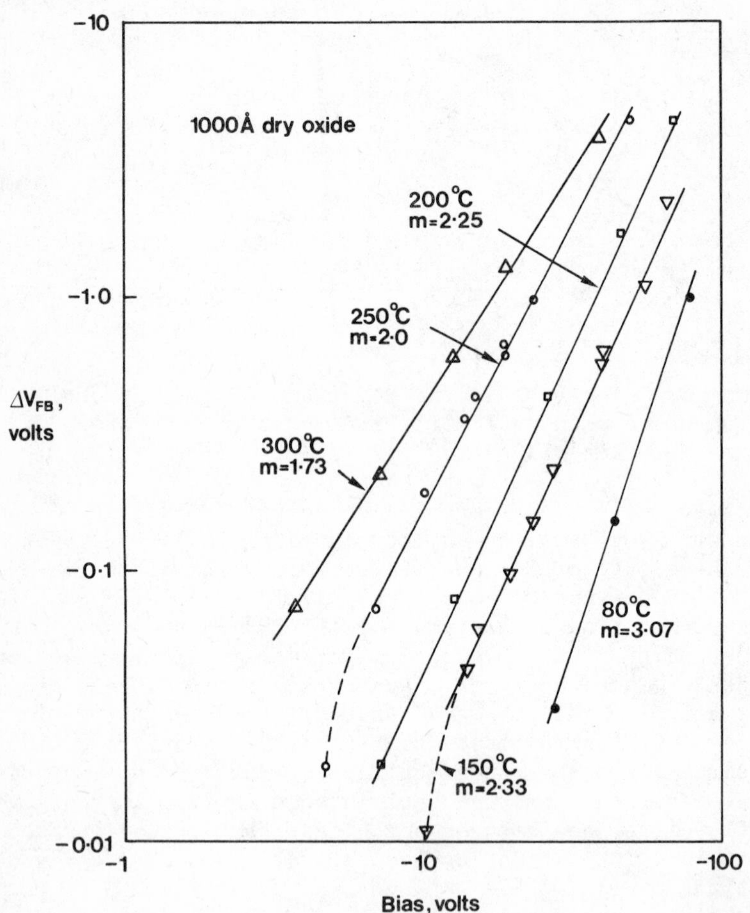

Figure 6.2 Variation of the hole trapping induced flatband shift as a function of applied voltage, at different temperatures. [After Sinha and Smith [140]).

The threshold voltage shift caused by trapped holes may be annealed by heating with the gate grounded, but the recovery is slower than the initial drift, and is also incomplete. Woods and Williams [65] quote an activation energy of 1.3 eV for this process. The holes may also be annihilated by injected electrons, but they cannot be detrapped with photons with energies of 4-5 eV [154].

6.3 Nature of the Trapping Mechanism

6.3.1 Methods of Trapping Holes

Experiments to measure the distributions of the hole traps in space and energy, and to find their capture cross-sections, only rarely make use of the long-term B-T stressing techniques used for accelerated life testing. Instead, one of three methods is commonly used to create holes in the oxide which may subsequently be trapped. Two of these methods involve photo-excitation of carriers.

Photons with energies greater than the bandgap of silicon dioxide create electron-hole pairs, and for energies in the 16 eV region, as are commonly used in this vacuum ultra-violet (VUV) technique, the light is adsorbed within 200Å of the (transparent) gate electrode. With the application of a positive voltage, the electrons are rapidly collected by the gate, whilst the holes drift to the silicon interface, where they are either trapped or annihilated by recombination with electrons from the silicon. This method has been employed by DiMaria et al [150], Powell [64] and Holmes-Siedle and Groombridge [151], amongst others. An anomalous build-up of positive interfacial charge with VUV illumination and negative gate bias has recently been reported by Weinberg and his co-workers [152, 153], and tentatively attributed to the motion of neutral exitons through the oxide.

The second photo-excitation technique uses much lower energy photons (~4 eV) to create hot holes in the silicon valence band. With a negative applied field, these holes are injected into the oxide valence band, and are then captured by the traps. Use of this technique has been described by Ning [63].

The third method of creating trapped holes is simply to apply very large negative fields. This is often [90a, 150] performed on a normal MOS capacitor, but other workers (Woods and Williams [65], Weinberg et al [149] and Iwanatsu and Tarui [155]) have dispensed with the metal electrode and applied the field by means of a corona discharge in air from an electrode suspended above the oxide. This method has the advantage of limiting the localized current density at any defects in the oxide, so large fields may be applied without causing catastrophic breakdown. A further advantage is that the slow decay of the charge when the corona is stopped may be monitored on an essentially uniform oxide, yielding further information on the traps.

6.3.2 Trap Densities

The three methods of trap-filling just described all lead to a very large amount of trapped charge; much larger, for example, than the amount responsible for the saturated threshold voltage shifts observed in accelerated life tests. This may be explained by stating that in the latter case only a small proportion of the hole traps are ever filled, possibly because some of them are located too deeply in the oxide, and are screened by holes already trapped nearer to the interface. When holes are present in the oxide as a result of very high fields or photo-excitation, however, the only limit to the trap-filling is electrostatic, that is, at very high densities of trapped charge the probability of "back-tunneling" into the silicon becomes high enough to preve[nt] any further net filling.

An alternative explanation for the different magnitudes of trapped charge found in life-test and "high stress" experiments is that in the latter case new traps are actually created by the stressing, corresponding to the linearly time-dependent high-field life-test mechanism.

Woods and Williams [65] have found the maximum value of the trapped charge N_{ot} to be 4×10^{12} charges/cm^2, and the results of Ning [63] imply a value of 1.4×10^{13} charges/cm^2. However, whereas the former authors anticipate the total number of neutral trapping centres to be in the 10^{15-16}/cm^2 region (based on the initial trapping efficiency and the capture cross-section), Ning's value of the cross-section seems to imply that all the traps are positively charged in saturation. The reason for this disparity is not clear, but might be the result of a field dependence of the cross-section.

6.3.3 Location of the Trapped Charge

The position of the trapped charge is known to be close to the Si/SiO_2 interface. Etch-back experiments [65] suggeste[d] that the charge centroid in corona charging experiments was ~130Å from the interface, but it was also suggested that the maximum in the density of the uncharged traps was much closer than this. More recently, DiMaria et al [146] have used an elegant technique to locate the charge centroid within 50Å of the silicon and, incidentally, to find some trapped holes very close to the gate electrode. This technique is based on changes to the photo I-V characteristics (with both polarities of gate bias) induced by the trapped charge. Whereas the effe[ct] of trapped charge on the C-V characteristics is a maximum when it is located right at the interface, its effect on the photo I-V characteristics depends on it increasing the interfacial electric field, and hence is a maximum when it is located a few tens of angstroms into the oxide.

The field-dependence of the flatband shift during B-T stressing has been fitted to (different) trapping models by Hofstein [141] and, more recently, by Jeppson and Svensson [90a], allowing values of 2.5Å and 3.2Å respectively to be obtained for the distance between the effective traps and the interface. These values are very similar to the Si-O distance of 3.1Å in SiO_2, and support the view that there is a sharp transition from silicon to oxide, with the hole traps caused by defects in the interface.

In general, energetically deep hole traps (in which the holes are relatively stable against detrapping) are only found at the interfaces, and not in the bulk. This is shown by the experiments of Powell [64], who found that holes introduced near the gate drifted almost the complete thickness of the oxide before being semi-permanently trapped. Deep traps can, however, be created in the bulk of the oxide by some processing steps such as, for example, implanting As or P [156]. Energetically shallow hole traps are thought to exist uniformly throughout the bulk of the oxide, and to account for the low mobility of holes. Curtis and Srour [157] have developed a multiple trapping model for hole transport which assumes a continuum of traps with a density decreasing exponentially from the oxide valence band edge, and which accounts for the transport properties at low temperatures. At room temperature, these traps are rapidly depopulated, the holes moving into the deeper interfacial traps [158].

6.3.4 Trapping Efficiencies and Capture Cross-sections

The hole trapping efficiency η_{eff} is simply related to the capture cross-section σ and the density of neutral traps N_o (per unit area) by

$$\eta_{eff} = \sigma N_o \qquad (6.4)$$

As the traps are filled, the trapping efficiency decreases, allowing the trap density and the capture cross-section to be calculated independently from the kinetics of the flatband shift, if the number of injected holes and the charge centroid are known. For the case of optically induced hot holes injected by low negative fields, Ning [63] obtained an initial trapping efficiency of 98.7%, and a capture cross-section of 3.1×10^{-13} cm^2. He assumed that the traps were distributed uniformly in the oxide, which gave a density of 1.4×10^{18} cm^{-3}

for the 1000Å thick oxide.

Powell [64] has found lower values of trapping efficiency for holes introduced near the outer electrode by a short (10 second) exposure to 10.2 eV photons, and then drifted to the Si/SiO_2 interface by positive bias. The hole photocurrent during the exposure, and the flatband shift resulting from it, are shown in Figure 6.3 as a function of the interfacial electric field. As the field is increased from zero, the hole

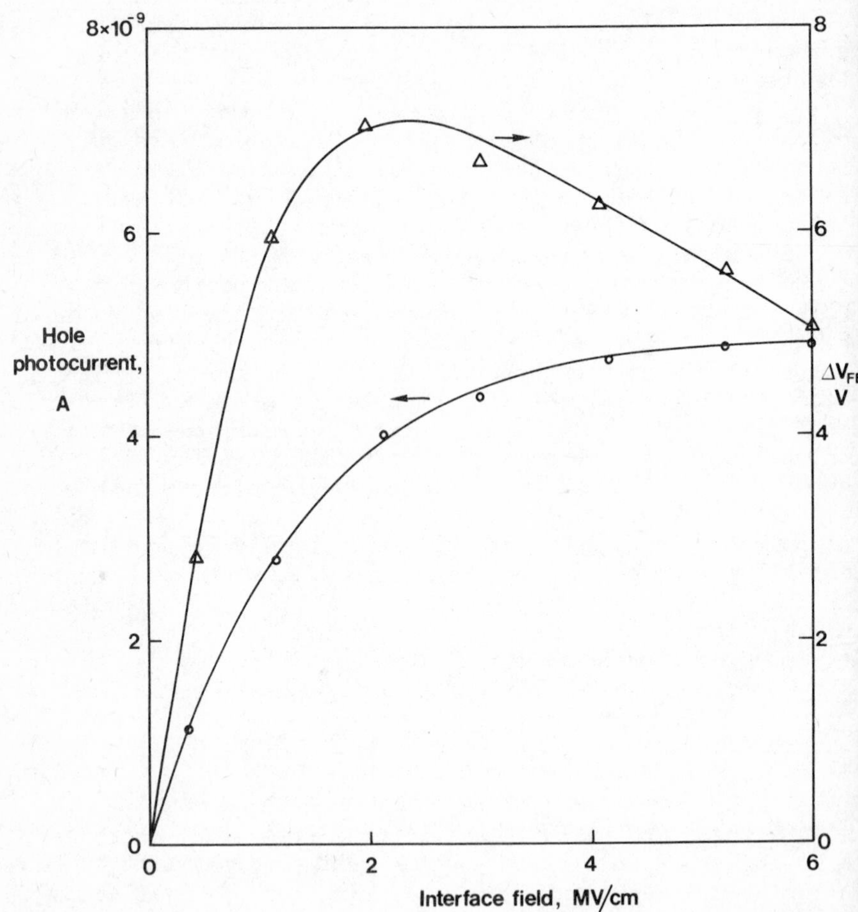

Figure 6.3 Field dependence of the hole photocurrent and flat-band voltage shift caused by a short exposure (10 seconds) to VUV (10.2 eV) photons, with a flux of 3 x 10^{11} cm^{-2} s^{-1}. Each point is a new device. (After Powell [64]).

current increases rapidly because of a reduction in the recombination probability, but it eventually saturates when all the holes are collected. The flatband shift (which reflects the total number of holes trapped at the interface during the exposure) initially increases with the hole current, but falls again above about 2 MV/cm, either because of annihilation by tunneling electrons, or by a reduction of the trap cross-section with high fields. Interestingly, the trapped holes increase the interfacial field to such a degree that the oxide current is increased (by up to two orders of magnitude) above the level corresponding to 100% quantum efficiency, due to Fowler-Nordheim injection of electrons. Values of the effective hole collection efficiency which can be extracted from Powell's data range from 53% at the lowest fields to 14% at the highest; the capture cross-section is again of the order of 10^{-13} cm^2.

6.4 Trapping Mechanisms

There are a number of possible mechanisms by which positive charge may be created by high negative fields, and the most likely are illustrated schematically in Figure 6.4. Process (1) is the direct tunneling of an electron from a neutral trap into the oxide conduction band. Although Woods and Williams [65] believe that this is the dominant mechanism at very high fields, the barrier is likely to be too wide for this mechanism to be significant during ordinary life tests, if the trap is deep enough to be stable against room temperature depopulation.

Process (2) shows the creation of an electron/hole pair by impact ionization by a hot electron, originally injected at the outer electrode. The hole could then be trapped, and the electron simply adds to the observed oxide current. Because of the wide bandgap of SiO_2 (~9 eV), only a small percentage of the electrons will gain the ionization energy, and the process is inefficient because the applied field causes the holes to drift away from the deep traps at the Si/SiO_2 interface. Nevertheless, impact ionization is often cited [159] as the mechanism for intrinsic dielectric breakdown at very high electric fields, although in this case the buildup of a quasi-stationary positive charge near the cathode is likely to be more important than the deeply trapped holes.

Process (3) represents the direct tunneling of an electron from a neutral trap into the silicon conduction band, without entering the oxide conduction band. This mechanism is critically dependent on the exact energy level of the trap, and is also limited to traps very close to the interface. Traps filled in this way could easily empty when the negative bias is

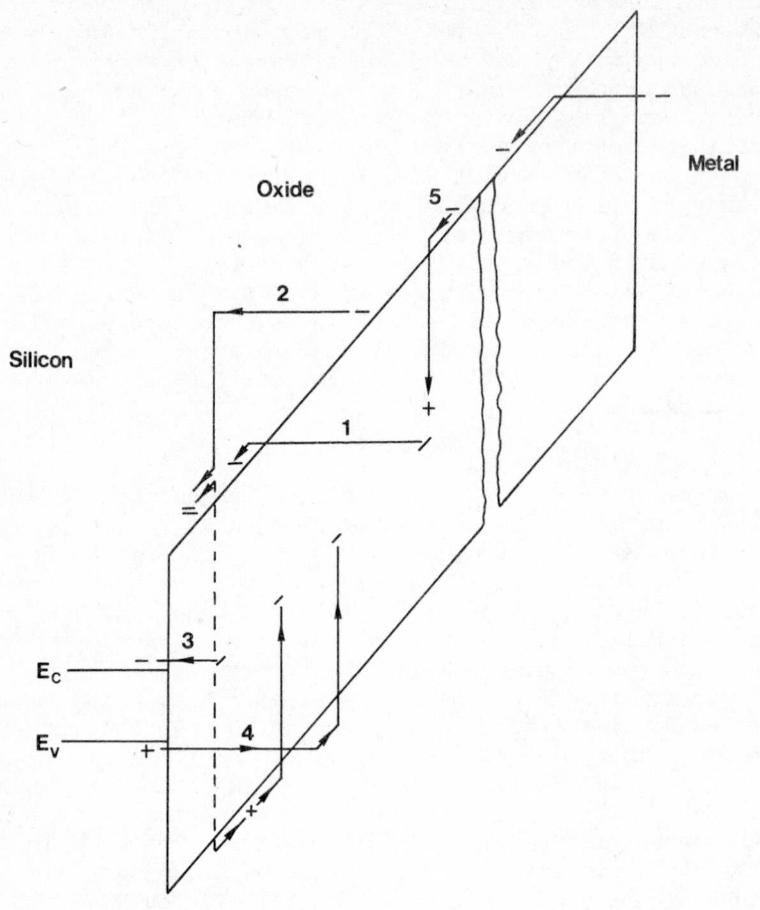

Figure 6.4 Band diagram illustrating the possible mechanisms of creating trapped holes with large negative fields. The various processes are described in the text.

removed. Although partial de-trapping of this type has been observed by Weinberg et al [149] and by Woods and Williams [65] it is thought that this mechanism of hole trapping is not the most significant during B-T life-tests.

The mechanism shown by process (4) is very similar to that originally proposed by Hofstein [141], and is the most easily associated with the term "slow hole trapping". It is the tunneling of holes from the silicon valence band into the SiO_2 valence band, from where they are trapped. The barrier height for this process is 4.7 eV, a very high value, leading Hofstein to propose that the traps, which are of the coulombic, attractive type, are located very close to the interface (as shown in Figure 6.5) so that their potential wells penetrate the silicon, thus making them accessible to holes with only moderate (~1 eV) thermal energies. One objection to this model is that experimentally a sharply defined

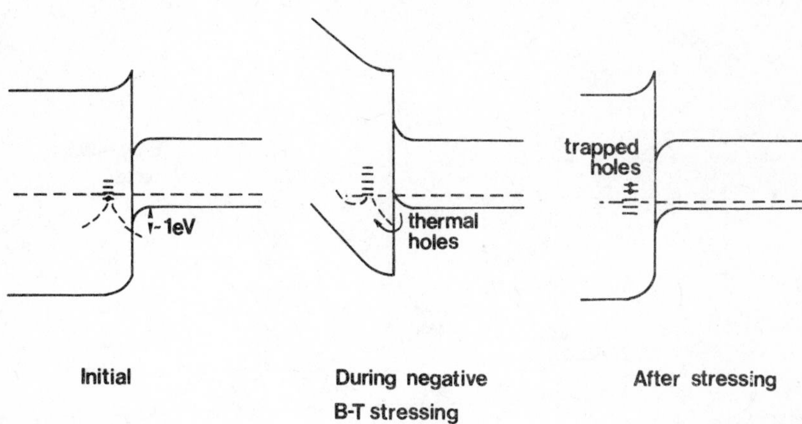

Figure 6.5 Hofstein's [141] model for slow hole trapping. The potential energy well of the hole trap extends into the silicon valence band, reducing the barrier height for thermal holes

thermal activation energy is observed, implying only a small range of trap positions and energies, which seems unlikely. It is sometimes thought that there is no hole injection into the oxide because of the absence of a steady-state hole current (established [149] by using a shallow p-n junction beneath the oxide to separate the carriers), but because of the high value of the trapping efficiency it is possible that the tunneling holes would not contribute to the oxide current.

Weinberg et al [149] have modelled the shift of the flat band voltage with time during negative bias stressing by assuming that the hole current can be represented by the simplified Fowler-Nordheim expression

$$J(t) = B \exp(-\beta_\lambda / E_o(t)) \qquad (6.5)$$

where B and β_λ are constants and $E_o(t)$ is the field in the oxide at the silicon interface. They found that the experimentally determined value of B is three orders of magnitude smaller than would be expected for a barrier height of 4.7 eV, but it would be consistent with the tunneling of hot holes, as depicted in Figure 6.6. These hot holes could be expected to be generated by impact ionization by electrons

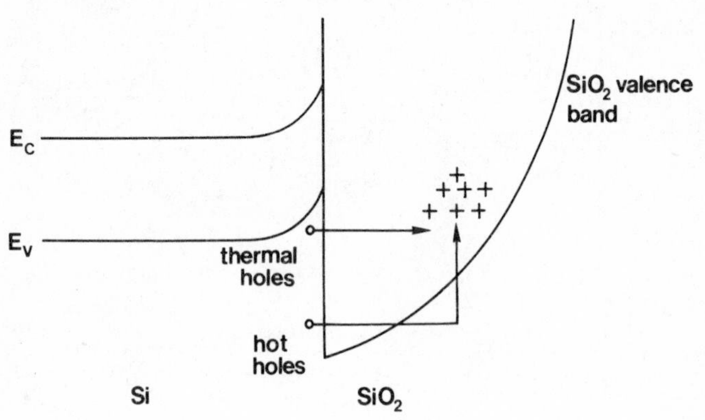

Figure 6.6 Two methods for hole tunneling from the silicon valence band, with subsequent distortion of the oxide field. (After Weinberg et al [149]).

entering the silicon from the oxide conduction band. The relatively weak voltage dependence of the flatband shift described in Section 6.2 is not properly understood, but may be a result of the field dependence of the capture cross-section, or due to neutralization of the trapped holes by conducting electrons at high fields, a process which will oppose the increased hole injection.

At very high negative fields (in the region where ΔV_{FB} increases linearly with time) Jeppson and Svensson [90a] propose that holes tunnel from the edge of the silicon valence band directly into the traps, as shown in Figure 6.7. From the field dependence of the trapping rate, they locate the trapping level 6.2 eV below the oxide conduction band, but it is difficult to reconcile this model with the fact that, after annealing, the traps refill much more quickly than during the original stressing. This result seem to indicate that new traps have been created by the action of the high fields.

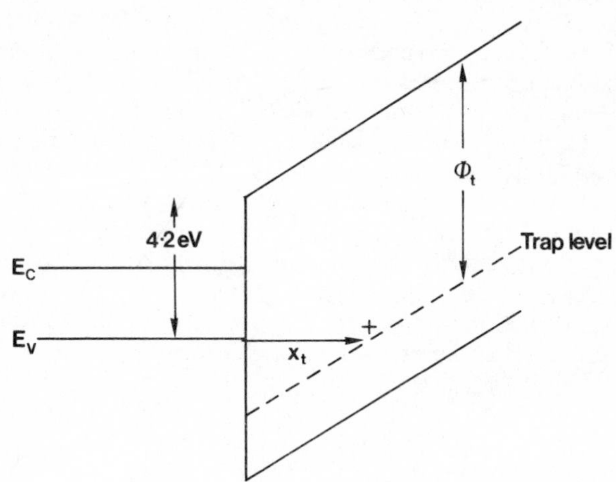

Figure 6.7 Tunneling model for hole trapping under very high negative fields, as proposed by Jeppson and Svensson [90a].

6.5 Origins of Hole Traps, and Processing Effects

Very much less has been written on the origins of hole traps compared to the descriptions of their properties, and even those models which have been proposed are largely unsubstantiated. Revesz [21], who gives a comprehensive review of the chemical and physical properties of silicon dioxide, considers hole trapping to be part of the intrinsic nature of the Si-O bond. There are two major objections to this view, however. Firstly, it would be equally applicable to the Si-O bonds in the bulk of the oxide as well as to those at the Si/SiO_2 interface, whereas it is now well established that almost none of the deep hole traps are located in the bulk. Secondly, the hole trapping drift saturates at values which are much too low if it is postulated that every Si-O bond is a potential trap. It is possible, however, that the breaking of strained Si-O bonds is responsible for the enhanced hole trapping observed at very high fields, and where the trapped charge increases approximately linearly with the total charge flow through the oxide.

Most models of hole traps are based on interface defects of some type, with the oxygen vacancy figuring prominently and corresponding to the observations of excess silicon in the oxide within a few tens of angstroms of the interface. Woods and Williams [65] associate hole traps with the E_1' colour centre (absorbance at 2120Å) observed in bulk silica after heavy irradiation. For this centre, it is proposed that an oxygen atom is first removed from the tetrahedral structure, with two silicon atoms forming a Si-Si bond across the vacancy but remaining approximately in their original positions This centre thus forms a neutral hole trap, becoming positively charged by loosing one of the non-bonding electrons. On becoming charged, there is an atomic re-arrangement of the centre from a tetrahedral to a planar arrangement, accounting for the difference in the activation energies of the trapping and subsequent annealing. According to the description of the E_1' centre given by Lell et al [160], its formation may be blocked by the presence of an OH group, as is shown in Figure 6.8. Since the presence of hole traps and surface state

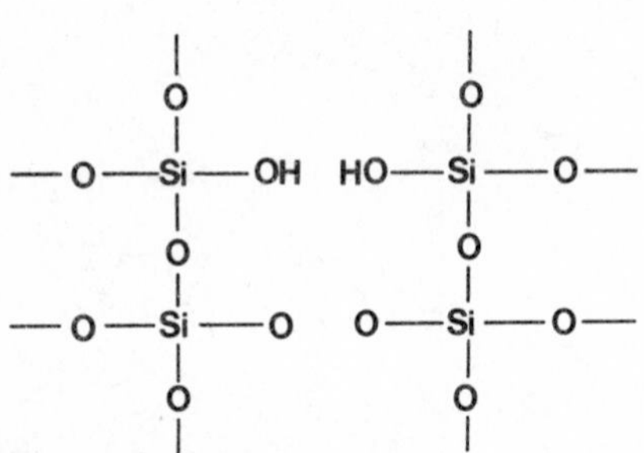

Figure 6.8 Blocking of the E_1' trapping centre by OH groups. (After Lell et al [160]).

are closely linked, this mechanism can explain the observation that wet grown oxides have lower surface states densities than dry ones before annealing, but that given appropriate annealing after growth, dry oxides generally prove to be more stable under negative B-T treatment. In this model, it is assumed that the OH group is removed relatively easily by stressing, leaving the original E_1' centre, but that in the annealed dry oxides the centres are completely irradicated.

The basic neutral defect in the hole trapping model of Jeppson and Svensson [90a] is a Si-H group at the interface. Under the influence of negative B-T stressing at moderate fields, this group reacts with an adjacent Si-O group to form a surface state, a positively charged interfacial silicon ion, an electron which is transported to the silicon, and an

Si-OH group. This reaction is illustrated schematically in the two-dimensional model of Figure 6.9. For the trapped charge to be stable, the OH group must diffuse away from the

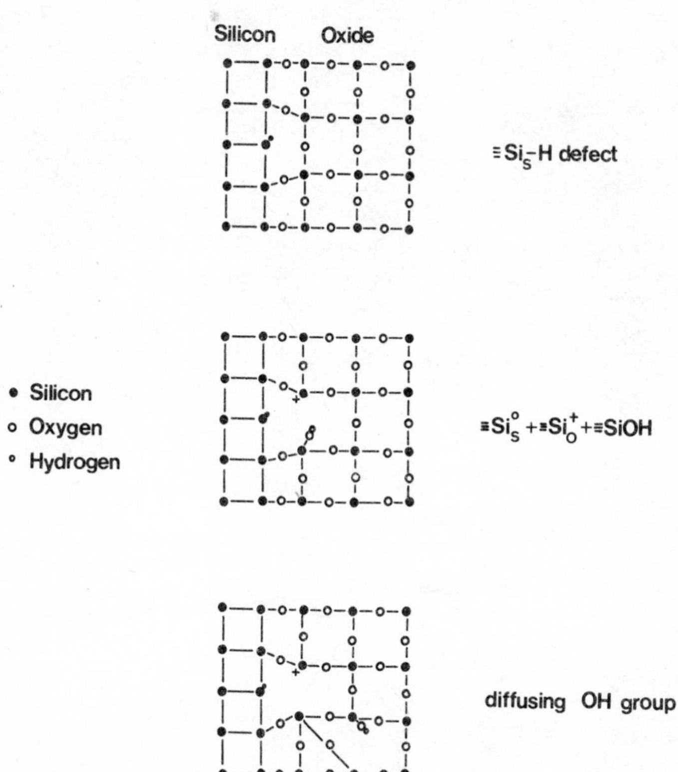

Figure 6.9 Schematic representation of the formation of a trapped hole and a surface state, as proposed by Jeppson and Svensson [90a].

vicinity of the defect to the bulk, a process which rate-limiting, and leads to the $t^{\frac{1}{4}}$ time dependence. Some of the OH groups may be trapped at other defect sites in the bulk, which accounts for the recovery during later annealing treatment being incomplete.

The dependencies of the density of hole traps (measured using the negative corona charging technique) on the oxide growth conditions and annealing treatment have been reported by Iwamatsu and Tarai [155]. They found that the density was independent of the silicon crystal orientation, in opposition to the normally held view [161] that it is proportional to the initial fixed charge density, and hence is higher for (111) silicon than for (100). The cause of this discrepancy is not known, but may be due to the very high fields created by the corona method. The response of the density of hole traps to post-oxidation and post-metallization annealing in hydrogen or nitrogen is at present unclear, but it has been shown [142, 155, 161] that oxide growth in HCl-containing atmospheres results in a relatively large build-up of positive interfacial charge during subsequent negative B-T stressing. As in the case of "clean" oxides, this positive charge build-up is accompanied by an increase in the surface state density, and has been attributed by Hess [161] to the straining of either Si-O or Si-Si (latent defect) bonds, caused by the presence of chlorine at the interface.

7

Electron Trapping

7.1 Introduction

Electron trapping is, in many regards, the complement of the slow hole trapping mechanism discussed in the previous chapter. With positive applied bias, electrons are injected from the silicon substrate into the oxide conduction band, where most drift quickly to the gate electrode (the electron mobility is approximately 20 cm^2/Vs [162]) and hence constitute a small gate leakage current. A small percentage of the injected electrons are captured by traps in the oxide, producing a positive shift of the threshold and flat-band voltages. Although the barrier for electron injection is significantly smaller than that for hole injection, resulting in the electron current with positive bias being much larger than the equivalent hole current with negative bias, silicon dioxide is virtually transparent to electrons (less than 1 in 10^5 electrons are trapped), whereas the trapping efficiency for holes approaches unity. Thus the electron trapping instability in n-channel transistors (for which positive gate voltage is the normal operating condition) is normally much smaller than that due to hole trapping in p-channel ones, at least where the insulator is pure thermal SiO_2. Indeed, in early transistors, the electron trapping instability was completely obscured by mobile ion drift, which also occurs with positive bias but produces a negative threshold voltage shift.

For all the instability mechanisms discussed in the previous chapters, the worst-case operating conditions for an MOS transistor are elevated temperature and high gate-voltage, with the source and drain grounded; these mechanisms may thus be investigated using an MOS capacitor as the test vehicle. For the case of electron trapping, however, normal transistor operation can give rise to a population of energetic ("hot") electrons in the silicon, which leads to an increased gate current and hence to larger instabilities. There are a number of possible means by which hot electrons may be created. One obvious source is avalanche breakdown of the source and drain p-n junctions, but this is not normally an allowed mode of operation in MOS circuits. Ning et al [163] have found that electrons created in the depletion regions of a transistor, and which are part of the normal thermal leakage current, can be accelerated by the electric field in the channel region towards the silicon surface, gaining sufficient energy to be emitted over the barrier into the oxide. The mechanism for generating hot electrons which is most likely to cause instability problems in modern, short channel transistors is the acceleration of the electrons which form the normal channel current (when the transistor is in its "on" or conducting state) by the field along the channel created by the drain voltage. This mechanism is a strong function of the geometry of the transistor, and also depends on the gate, drain and substrate voltages in a complex manner.

As well as producing carriers for injection and subsequent trapping in the oxide, thus producing threshold voltage shifts hot electron effects also give rise to instabilities in several other transistor parameters. For example, there is often an increase in the leakage current of the drain junction [164] due to crystallographic damage in the drain depletion region, and the gain factor β is usually observed [165] to decrease after a period of stressing in conditions designed to produce hot electrons. In addition, the presence of hot electrons in the substrate gives rise to a range of second-order effects on transistor operation; one such effect is the initiation of parasitic lateral n-p-n bipolar transistor action [166], which reduces the source to drain punch-through voltage and produces a potentially destructive negative resistance region in the I_D-V_{DS} characteristics. Another effect arises in dynamic MOS memory circuits, where hot electrons from the channel of a conducting transistor can drift to an adjacent memory cell and cause a loss of data [167]. These effects occur in new, unstressed transistors, and they become more pronounced after extended periods of operation, particularly with high applied voltages.

Hot electron trapping proceeds by two independent stages. The first stage is the generation of the hot carriers in the substrate, their transport to the silicon surface, and their injection into the oxide. The resulting gate current is determined by the biasing conditions, the temperature, the doping profile in the substrate and the depth of the source and drain junctions, and by the channel length. The second stage is the capture of a small percentage of the injected electrons by oxide traps. The density and capture cross-sections of these traps are determined principally by the processing technology used for growing the oxide, but the trapping efficiency also depends on the temperature and on the electric field in the oxide. Once trapped, the electrons alter the emission probability for further injection, with the result that the maximum amount of charge trapped by extended periods of stressing depends on the initial value of the gate current as well as on the total number of traps present. The most obvious effect of the trapped charge on the transistor parameters is an increase in the threshold voltage, but it also alters the breakdown voltages of the junctions [164] and, because the charge is non-uniformly distributed along the channel, it has a secondary effect on the gain factor.

It can be seen that hot electron induced instabilities are considerably more complex than those dealt with in the earlier chapters, and although the basic mechanisms of degradation are well understood it is not possible at present to model the changes during life of the parameters of an arbitrarily chosen transistor as a function of its operating conditions.

7.2 Reliability Effects

7.2.1 "Cold" Electron Trapping

Trapping in MOS capacitors, or in transistors with gate bias only, is limited by Fowler-Nordheim tunneling of electrons into the oxide conduction band [168]. (Weinberg [169] has noted that F-N tunneling is not strictly accurate for injection from silicon (as opposed to a metal), but his alternative model gives qualitatively the same currents). At low stress levels, this current produces a small (typically 20-100 mV) and rapidly saturating positive shift of threshold voltage as the relatively small numbers of electron traps with large capture cross-sections are filled.

With very high fields (> 8.5 MV/cm), or moderate fields (5-8 MV/cm) and high temperatures (200°C), a different behaviour is observed; the threshold voltage increases linearly with time, and shifts of several volts are observed before dielectric breakdown occurs and the oxide fails catastrophically. These results have been interpreted by Solomon [170], who used a ramped I-V technique, as being due to the filling of traps with small ($< 10^{-18}$ cm^2) cross-sections situated near to the Si/SiO_2 interface. These traps are very numerous but, because of their small cross-section, significant filling only occurs when the gate current is large. An alternative explanation is due to Harari [171, 172], who maintained a constant gate curre during stressing by increasing the gate voltage as the trapping progressed. He suggests that new traps are actually created near the interface by the breaking of Si-O bonds. Thi is a more likely explanation as it agrees with the observation [171] that although the traps can be emptied by negative bias, they subsequently refill at smaller positive voltages than during the initial stressing.

Electron injection and trapping from the gate electrode of poly Si-oxide-silicon samples has been observed by Hickmott [173] after only a few minutes with negative gate bias at room temperature. The charge was located ~ 30Å from the gate interface, which was sufficient to give shifts in the flat-band voltage of over 200 mV, as the oxide thickness was only 370Å. The magnitude of the shifts increased with increasing temperature, but saturated at about 60°C, suggestin that the traps involved are located just below the oxide conduction band; this is confirmed by their relaxation behaviour with positive applied voltage. Although the average applied fields in Hickmott's experiments were not large (< 4 MV/cm) it is probable that defects in the poly-Si/SiO_2 interface gave rise to much larger local fields, so producing larger oxide currents than would normally be expected, as is discussed in more detail in the next chapter.

7.2.2 Channel Electron Effects

Electrons travelling from the source to the drain in the channel of a conducting n-MOST gain energy from the longitudinal electric field, and can loose it again in collisions with the semiconductor lattice. The most important energy loss mechanism is optical phonon scattering, with an energy of 0.063 eV and a mean free path of approximately 90Å at room temperature [165]. For fields in excess of ~100 kV/cm the electrons can no longer remain in equilibrium with the lattice, and are heated by the field. Those electrons which gain the ionization energy (3/2 E_G) can create electron/hole pairs in ionizing collisions; the secondary electrons merely augment the drain current, causing a deviation from the predictions of simple transistor models, but the holes drift to the substrate, forming a parasitic substrate current. This substrate current is plotted in Figure 7.1 as a function of th gate voltage, and displays a characteristic bell-shaped curve. The rising section of this curve is due to the quadratically increasing drain current as the transistor is turned harder on and the falling section is caused by a quasi-exponential decre in the efficiency with which the channel electrons are accelerated, resulting from a redistribution of the longitudinal field along the channel. After the transistor ha

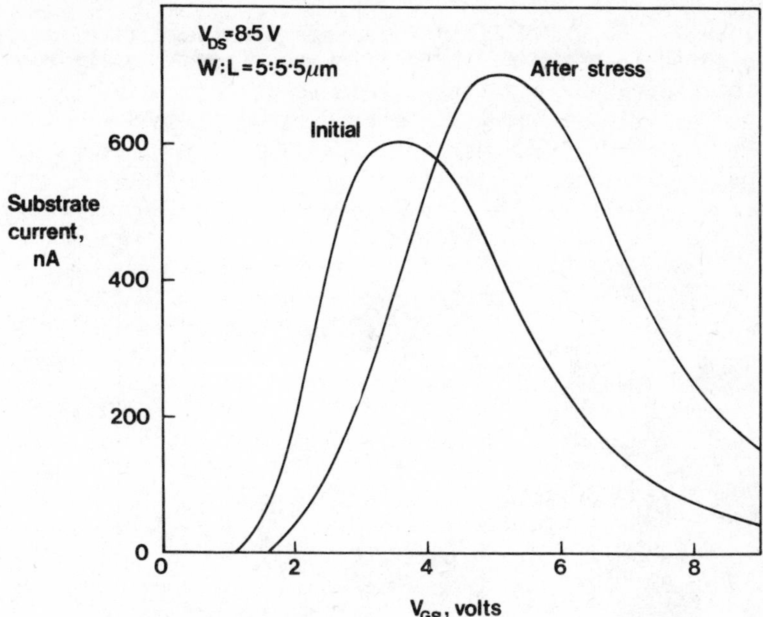

Figure 7.1 Substrate current of an n-channel MOST before and after hot electron injection.

been stressed to induce electron trapping, the maximum substrate current increases, and also occurs at a higher gate voltage, due to the further redistribution of the longitudinal field produced by the trapped charge. This increase in the unwanted substrate current results in several reliability hazards. Firstly, for integrated circuits with on-chip substrate bias generation, which have a limited current-sinking capability, the current may result in circuit malfunction. Secondly, the voltage drop caused by the current flowing through the substrate spreading resistance can forward bias the source junction, initiating lateral n-p-n action, and thirdly the holes can be accelerated in the depletion region and undergo secondary ionizing collisions. Although most of the resulting electrons are again collected by the drain, some escape to the neutral bulk where they can diffuse quite long distances (several hundreds of microns) and interfere with the operation of other parts of the circuit.

Channel electrons which avoid ionizing collisions continue to gain energy from the longitudinal field, and can eventually become hot enough (>3 eV) to be injected over the Si/SiO_2 interface barrier if they are scattered towards the surface by a phonon collision. The injection efficiency κ, defined as the ratio of the gate current to the drain current, is shown in Figure 7.2 as a function of the drain voltage at a number of ambient temperatures. The efficiency increases rapidly with increasing drain voltage (because of the increase in longitudinal field) until the drain voltage exceeds that

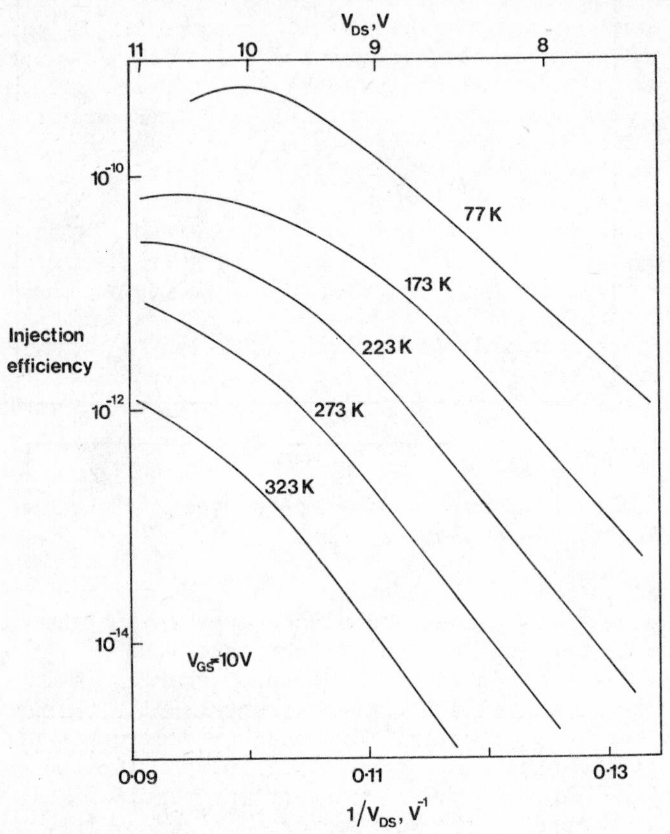

Figure 7.2 Dependence of the injection efficiency of a 5 x 5 µm transistor with drain voltage and ambient temperature. V_{GS} = 10V.

of the gate, when it saturates or exhibits a peak. One reason for this peak is that, when the transistor is operating in the saturated region, the lateral field in the silicon opposes the drift of hot electrons created deep in the drain depletion region towards the surface. Also, some of the electrons which are injected into the oxide are returned to the drain and do not form part of the measured gate current, although they are subject to trapping at sites near the interface. The worst case biasing conditions for use in accelerated life tests are usually taken to be $V_{GS} = V_{DS}$, with the source and substrate grounded. (Substrate bias has only a small effect on the gate current).

Figure 7.2 also shows that the injection efficiency increases at low ambient temperatures, due to the increase in the mean free path for phonon scattering [174] which allows the electrons to gain more energy between collisions. Since this same effect also causes an increase in the gain factor β, and hence in the drain current I_D, the gate current ($= \kappa I_D$) is increased dramatically by low temperature operation. Channel hot-electron induced degradation is thus unique amongst the instabilities considered in this book by requiring low temperatures for accelerated testing.

Because the longitudinal field in the channel of a conducting transistor is larger at the drain end of the channel than near the source, the hot electrons are created and trapped predominantly near the drain. Consequently, after a period of stressing, the transistor has a non-uniform "localized threshold voltage", giving rise to the transfer characteristics shown in Figure 7.3. In the saturated region, when the transistor is operated in the forward mode (with the

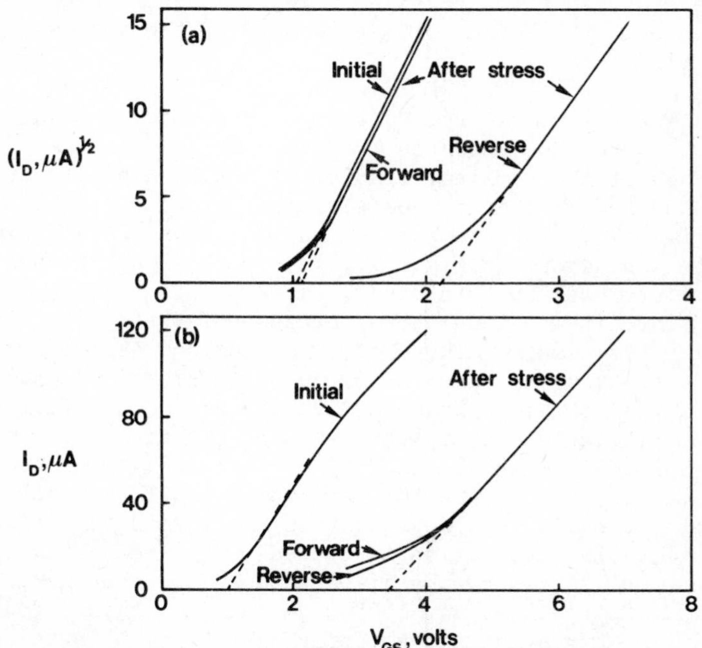

Figure 7.3 (a) Saturated and (b) unsaturated transfer characteristics of a short channel n-MOST before and after stressing at 77K.

source and drain connections the same for measurement and stressing) the trapped charge is located over the pinched-off end of the channel and hence has little affect on the characteristics, whereas in the reverse mode (that is, with the source and drain connections interchanged between stressing and measuring) the charge is over the beginning of the channel and so there is a large increase in the threshold voltage. In the unsaturated region, there is no pinched-off region, and so both the forward and the reverse curves degrade.

The presence of oxide charge at the drain end of the channel also produces changes in the gain factor, as the transistor can be considered as two devices with different threshold voltages in series, producing a shortening of the effective channel length, and hence an increase in β [132] (especially in the reverse mode). This effect is observed after moderate periods of stressing, but after prolonged periods at high voltages the gain reduces again, due to damage caused to the silicon lattice in the vicinity of the drain by collisions with the high energy electrons. This damage also results in an increase (sometimes catastrophic in the drain junction leakage current, as well as an increase in the drain breakdown voltage because of the increased scattering.

7.2.3 Time, Temperature, Voltage and Geometry Dependencies of Channel Hot Electron Trapping

Some of the earliest studies of the threshold voltage shifts caused by channel hot electrons are due to Abbas and Dockerty [175, 176], whose results are shown for a variety of gate insulators in Figure 7.4. For the long channel transistors used in this study, large shifts are only seen with oxide-nitride insulators, because of their high trapping efficiency. They modelled the log (ΔV_T) : log(t) dependence by using first order trapping kinetics, and assumed that the saturation of ΔV_T was due to the filling of all the available traps. Later authors [164, 165], who observed much larger shifts for short channel transistors with clean oxides, found that the saturation level, as well as the rate of shift, was dependent on the stress conditions, suggesting that saturation is caused by the reduction of the gate current caused by the field produced by the charge already trapped. Cottrell et al [165] have modelled this effect by assuming a Maxwellian hot-electron distribution in the silicon with an electron temperature of T_e and that the trappin efficiency η for electrons injected into the oxide is independent of the number already trapped (that is, the number of neutral traps is always much larger than the trapped charge density). They obtain

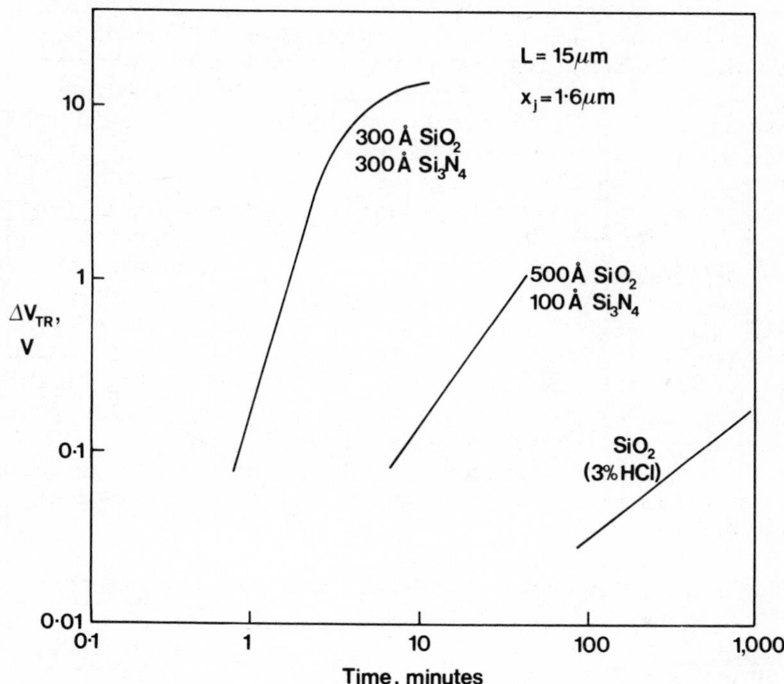

Figure 7.4 Reverse threshold voltage shifts as a function of time for n-channel MOSTs stressed with $V_{GS} = V_{DS} = 15V$, $V_{SX} = 3V$, and at room temperature. (After Abbas and Dockerty [176]).

$$Q_T = Q_o \ln (1 + t/\tau) \tag{7.1}$$

where
$$Q_o = CkT_e/q \tag{7.2}$$

and
$$\tau = Q_o/\eta I_o \tag{7.3}$$

Q_T is the trapped charge, C is the capacitance associated with filling the traps and I_o is the initial gate current (in the absence of any trapped charge). The effect of Q_T (which is located near the drain) on the transistor characteristics must be evaluated using a two dimensional solution to Poisson's equation, which predicts that the fractional decrease in the channel current is proportional to the square of the trapped charge, as is shown in Figure 7.5. Note that in this approach the effects due to changes in V_T and β are combined.

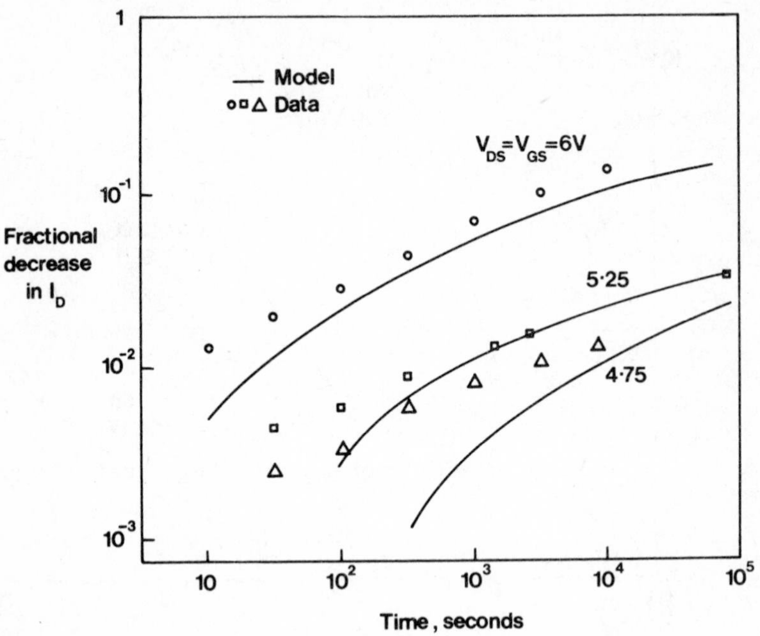

Figure 7.5 Fractional decrease in the channel current of a 1:11 μm transistor, characterized at $V_{GS} = V_T + 1V$, $V_{DS} = 0.5V$ and $V_{SX} = 1V$ (reverse mode), during stressing at various voltages. (After Cottrell et al [165]).

The dependence of the threshold voltage shift on the applied voltages can in principle be calculated from their effects on the initial gate current I_o and on the electron temperature T_e if the reduction in the trap capture cross-section with increasing oxide field is neglected. Because of the complexity of the three-dimensional generation and transport of hot electrons, simple empirical models have to be used for the voltage dependence of I_o. Euzent [177] used

$$I_o = KI_D \exp (a V_{DS}) \exp (b V_{GS}) \quad (7.4)$$

where K, a and b are constants, and the voltage dependence of the drain current must be considered separately. This equation holds approximately only for $V_{GS} < V_{DS}$, but most authors prefer to restrict the problem to the worst-case $V = V_{GS} = V_{DS}$ condition, where the relationship can be expressed empirically as [165]

$$I_o = K \exp\left[\left(\frac{L}{L_o}\right)^{-\gamma} \left(\frac{V}{V_o} - 1\right)\right] \tag{7.5}$$

where L is the channel length and K, L_o, γ and V_o are constants whose values depend on the doping profiles and the ambient temperature.

There is a lack of accurate experimental data on the rates of practical transistor threshold voltage shifts during life-testing, However, it can be stated that the shift after a fixed period of stressing increases approximately exponentially with the stress voltage and with the reciprocal of the channel length, and decreases exponentially with the absolute temperature of the silicon [175, 178]. (It should be noted that at very low temperatures the trap cross-sections may increase, and energetically shallow traps can capture electrons, producing further increases in the level of trapped charge [179].)

7.2.4 Leakage Hot Electrons

Electrons in the neutral bulk of the semiconductor diffuse to the depletion region beneath the transistor gate and then, together with thermally generated electrons from the depletion region, are accelerated towards the surface and emitted over the interface barrier. For transistors with vertical dimensions much smaller than their horizontal ones, this current flow is essentially one-dimensional, and so the charge is trapped uniformly over the gate area. If the gate voltage is large enough to invert the surface (and the source and drain are grounded to prevent any channel current) then the electrons are heated by the substrate bias, and the barrier height is obtained by considering the Schottky lowering produced by the gate voltage. In most practical cases the diffusion current component is much larger that the generation current one, but at high substrate bias there may also be some carrier multiplication. Models for this emission process are considered in Section 7.3 (for the one-dimensional case).

There have been relatively few studies of leakage-electron induced threshold voltage shifts, mainly because this process is not as significant as the channel electron one, especially for short channel transistors where many of the leakage hot electrons are swept to the source and drain diffusions, reducing the effect [165]. Some typical results obtained with oxide-nitride insulators by Ning et al [163] are shown in

Figure 7.6. The linear shift of V_T with time results from the constant emission current and the large numbers of

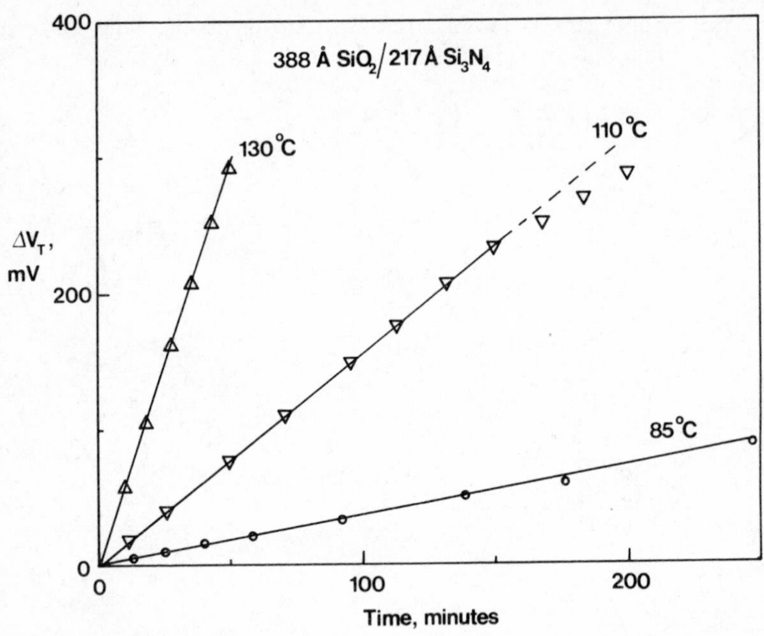

Figure 7.6 Threshold voltage shifts due to the trapping of leakage electrons in transistors stressed with V_{DS} = 0V, V_{GS} = 6.5V and V_{SX} = 10V. (After Ning et al [163]).

traps in the nitride. Despite the reduction of the mean free paths of electrons in the substrate with increasing temperatures, these results show a positive activation energy of 0.8 eV due to the increased supply of diffusion electrons. Qualitatively similar results have been obtained by Chaudhari [180] for transistors with phosphorous-doped gate oxides. He analysed these results using the rather curious assumption that the capture cross-section of the traps fell exponentially with the distance from the Si/SiO_2 interface, and obtained a $(\Delta V_T)^{\frac{1}{2}}$: log(time) relationship, and an activation energy of 1.06 eV.

7.2.5 Hot-electron Trapping in p-channel Transistors

The negative applied biases used to operate p-channel prevent the generation of hot electrons by either the channel current or leakage current mechanisms. (Hot holes can, in theory, be generated in these circumstances, but because of their low mobility and the high barrier for hole injection into the oxide, this effect is not normally observable.

For example, the slow hole trapping instability is observed without applying substrate bias, and is not enhanced when the transistor is conducting). Hot electrons can, however, be generated by avalanching the drain junction and, if the drain voltage is more negative than that of the gate, the oxide field in the drain overlap region aids electron injection. This process has been observed by Dunn and Mellor [181] in a variety of commercial transistors. The effect of trapped electrons on the parameters of a p-channel transistor is, of course, rather different than for the n-channel case. For example, the region of the channel near the drain now has a smaller (in magnitude) threshold voltage than elsewhere, but no channel current can flow until the source end inverts and so the total threshold voltage is, to a first approximation, unchanged. The variation of the "localized threshold voltage" does, however, produce an increase in the gain factor, and the breakdown voltage of the drain junction "walks out" to higher voltages.

Avalanche breakdown is normally avoided in integrated circuit operation, but Coe [182] has also observed hot electron effects in p-channel transistors with the drain biassed below breakdown, where there is still some avalanche multiplication. Using short-channel transistors (3-4 μm) he found that the fractional change in the effective channel length was large enough to produce a 15 mV reduction in the magnitude of the threshold voltage after only 10 seconds of stressing with a drain voltage 5V below breakdown.

7.3 Hot-electron Generation Methods for Electron Trapping Investigations, and Emission Models

The complex, three-dimensional nature of the channel hot electron mechanism makes it extremely difficult to model theoretically, and although Phillips et al [183] have described such a model they have ignored several important factors making it of little practical use in predicting the actual oxide current. Similarly, the injection of hot electrons produced by the surface avalanche breakdown of a gated p^+-n junction results in a highly non-uniform current [184].

For experimental determinations of the trapping properties of the oxides, a one-dimensional flow of electrons in the substrate, producing a uniform oxide current, is highly desirable; there are several methods by which this may be achieved. Nicollian et al [185] employed a large ac signal (to prevent surface inversion) of relatively high frequency applied to the gate electrode of a large-area MOS capacitor to drive the silicon into deep-depletion and then into avalanche breakdown. Hot electrons formed in the avalanche region are transported to the surface, and give rise to a pulsed oxide current. Nicollian, who used a sinusoidal signal, maintained the average "dc" current at a constant value by using feedback to increase the amplitude of the signal to compensate for the effect of the trapped electrons. Although the increase in the signal tracks the change in the flat-band voltage, Young et al [186], who used a squarewave, believe that it is more accurate to interrupt the avalanche periodically and make direct measurements of V_{FB}.

The avalanche injection technique is often used when the requirement is to produce a large, uniform oxide current, but for studies of the emission process itself it suffers from the fact that the electrons produced by impact ionization can have appreciable energies, and the applied bias controls both the oxide field (which in turn affects the interface barrier height and the capture cross-section of the traps) and the heating field in the substrate. This second disadvantage can be overcome by using a large-area MOS transistor (often with closed geometry) which allows the surface potential to be "pinned" at $2\psi_F$; the oxide and the substrate fields can then be determined independently by the gate voltage and the substrate voltage respectively. Two methods exist for introducing low-energy electrons into the substrate depletion region, both of which require some modification to the basic MOS processing technology. The first method, which has been used by Verwey et al [187] and Young [188], uses a forward biassed planar epitaxial p-n junction which underlies the n-channel transistor structure. The second method, described by Ning et al [174] dispences with the need for an epitaxial layer but instead requires the gate material to be transparent, normally achieved by ~100Å of gold or aluminium, or by several thousands of angstroms of poly Si. Electron/hole pairs can be created in this structure by photons from a simple tungsten filament, and the total electron flux incident on the interface is given by twice the value of the drain current if it is completely surrounded by the source (which also collects electrons generated outside the transistor region).

Early models [187] of the emission process made use of the "lucky electron" model, in which only those electrons which acquire sufficient energy to surmount the interface barrier between energy-loss collisions are injected. In the band diagram of Figure 7.7 only electrons travelling to the interface from distances greater than d can gain sufficient energy to surmount to barrier, and the probability of travelling this distance without an energy-loss collision is

$$P = \exp(-d/\lambda) \qquad (7.6)$$

where λ is the combined mean free path for phonon and

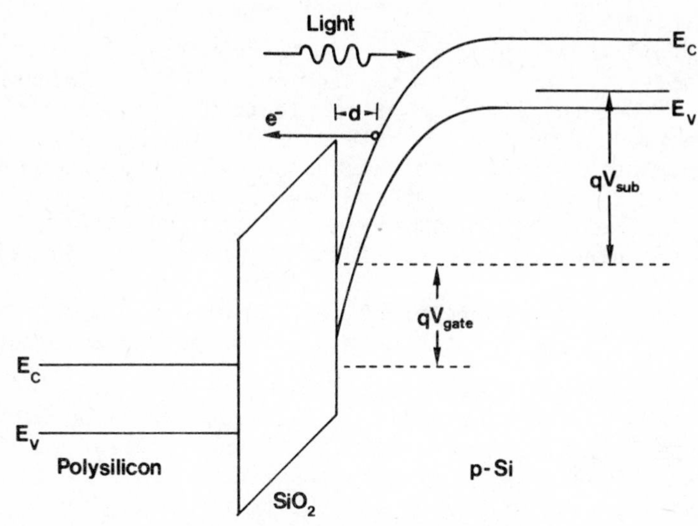

Figure 7.7 Schematic energy band diagram for hot electron injection

ionization collisions. The emission probability also turns out to have the form of Equation 7.6, where the distance d can be calculated from the doping profile in the substrate and the substrate bias if the barrier height is known. The barrier is lowered from its zero-field value of 3.1 eV by the Schottky-lowering produced by the oxide field and also, as described in a semi-empirical model by Ning et al [174], by a term which accounts for the tunneling probability of hot electrons arriving at the interface with insufficient energy to be emitted directly over the barrier. This second term is particularly important at low emission probabilities. The effective barrier height can thus be expressed as

$$\phi_b(\text{effective}) = 3.1 \text{ eV} - \underbrace{\beta E_{ox}^{\frac{1}{2}}}_{\text{barrier lowering}} - \underbrace{\alpha E_{ox}^{2/3}}_{\text{tunnelling effect}} \qquad (7.7)$$

where $\beta = 2.6 \times 10^{-4}$ e(V cm)$^{\frac{1}{2}}$ for SiO_2, and α is found by fitting to the experimental results.

This model can predict emission probabilities in good agreement with those observed experimentally but Troutman [198] has produced a much more sophisticated model which shows that the emitted electrons actually undergo many phonon collisions on their way to the surface. Thus the values of λ obtained from the lucky-electron model should not necessarily be regarded as the true values of the mean free path.

7.4 Electron Trapping

7.4.1 Trapping Theory

The most widely accepted model of electron trapping uses first-order kinetics, and is due to Ning and Yu [61]. Because the trapping efficiency (the number of electrons trapped for each one injected) is small, the gate current density j_G is constant throughout the oxide thickness, and also the oxide field may be taken as constant as long as the total threshold voltage shift is reltively small. Finally, it is assumed that the trapped electrons are not thermally re-emitted during the experiment.

For a trap with a capture cross-section of σ and a density of N_T, the rate equation for the density of trapped electrons n_T is;

$$\frac{dn_T}{dt} = \frac{j_G \sigma}{q} (N_T - n_T) \qquad (7.8)$$

which, for n_T $(t = 0) = 0$, has a solution

$$n_T(t) = N_T [1 - \exp(-\sigma N_{inj})] \qquad (7.9)$$

where N_{inj} is the number of electrons injected into the oxide and is obtained by integrating the current. The trapping efficiency η is determined by

$$\eta = \frac{q}{j_g} \int_0^d \frac{dn_T}{dt} .dx = N_{TT} \sigma \exp(-\sigma N_{inj}) \qquad (7.10)$$

where d is the oxide thickness and N_{TT} is the equivalent areal density of traps. The total trapped charge may be inferred from the change in the threshold voltage:

$$\Delta V_T = \frac{q}{C_{ox}} \cdot \frac{\bar{x}}{d} \int_0^d n_T(x).dx \qquad (7.11)$$

$\bar{x}$ is the centroid of the trap distribution, which in general is unknown, so it is convenient to introduce an effective trapping efficiency η_{eff} by

$$\eta_{eff} = \frac{d(C_{ox} \Delta V_T)}{dt} \cdot \frac{1}{j_G} \qquad (7.12)$$

and an effective trap concentration per unit area N_{TTeff}

$$N_{TTeff} = \frac{\bar{x}}{d} \cdot N_{TT} \tag{7.13}$$

so that

$$\eta_{eff} = N_{TTeff}\, \sigma \exp(-\sigma N_{inj}) \tag{7.14}$$

The difference between η and η_{eff} is determined by the spatial distribution of the traps; for a uniform distribution $\eta_{eff} = \frac{1}{2}\eta$, whereas $\eta_{eff} = \eta$ if all the traps are located at the Si/SiO_2 interface. (The literature often uses the symbol N_{ot} to represent the number of trapped charges per unit area, assuming them all to be at the interface).

Experimentally, the gate current and the threshold voltage are monitored as a function of time, allowing a plot of log (η_{eff}) against N_{inj} to be made, as is shown in Figure 7.8. Using Equation 7.14, σ and N_{TTeff} can be extracted from this plot from the slope and intercept respectively. If several trapping centres, with different cross-sections, exist, then the plot will consist of a number of linear segments, and the different values of σ can be extracted as long as they are sufficiently distinct. The larger traps fill at lower values of N_{inj} than the smaller ones, but the saturated shifts of the threshold voltage that they produce are dependent only on their density.

The centroid of the trapped distribution is normally [190-192] determined by the photo I-V technique, using electron injection from both electrodes, and this method can be combined with etch-back experiments [66, 193] to find the complete distribution of charge. For photo-depopulation experiments in which the electrons are not re-trapped, and for charging experiments in which the trapping efficiency is unity (ie for insulators with very high trap densities, such as Al_2O_3), the charge centroid can be determined directly from the flat-band shift [194].

7.4.2 Trap Parameters and Origins

In contrast to the straightforward trapping of holes near to the Si/SiO_2 interface, there is a wide range of distributions of electron traps with many different properties. The traps may be classified in several ways, for example by their charge state, their capture cross-section, their spatial and energy distributions, or by their physical origins. A

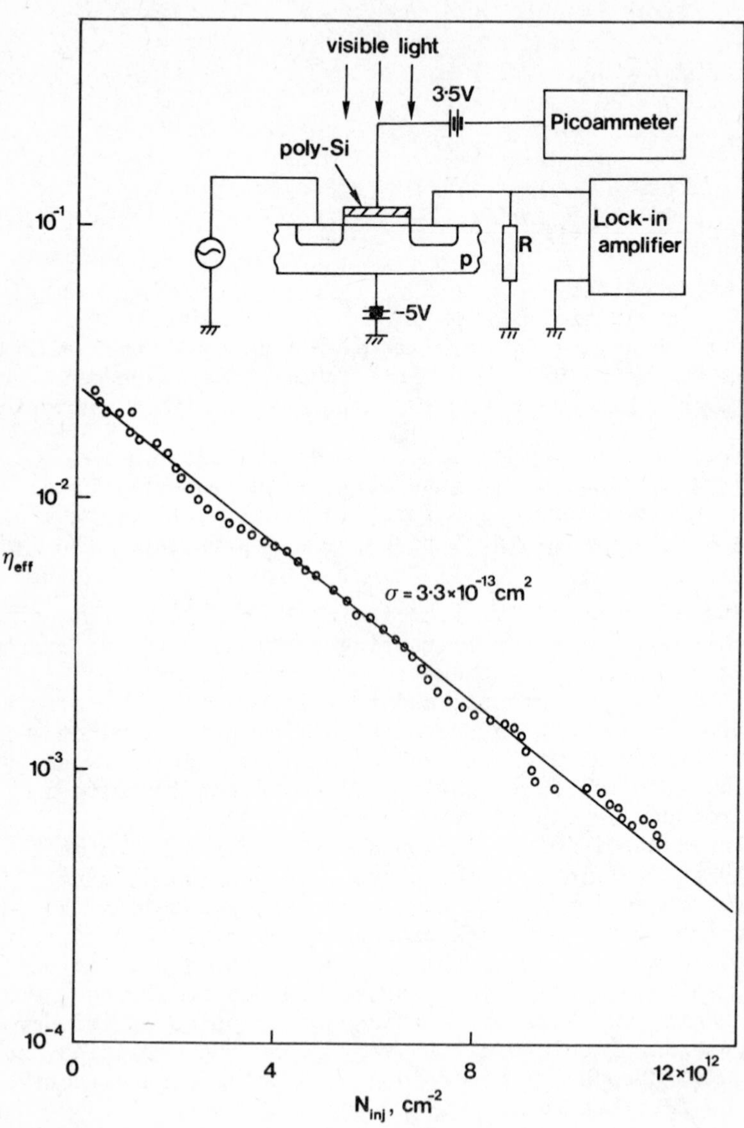

Figure 7.8 Semilog plot of the effective trapping efficiency vs total optically induced injected charge, at the relatively low injected charge density of 4×10^{-12} A (After Ning and Yu [61]).

summary of some of the traps observed by various workers is given in Table 7.1; although there is incomplete agreement on the exact properties of all the traps, a review of the current understanding has been given by DiMaria [154].

The largest trap has a capture cross-section of $\sim 10^{-13}$ cm^2, and it is widely observed [61, 62, 180, 196] in oxides without any intentional contamination. Arnett and Young [195] observed a very high density of similar traps in films of silicon nitride. These traps are Coulombic attractive centres, and the value of σ at zero fields is close to the theoretically predicted value of $10^{10}/\varepsilon_o\varepsilon_{ox}$ cm^2; however, they show a E_{ox}^{-3} field dependence rather that the expected $E_{ox}^{-3/2}$ one, possibly due to the increase in the electron temperatures. They are located near to the Si/SiO_2 interface (within 20Å according to Powell and Berglund [192] and usually have areal densities 10^{11}-10^{13} cm^{-2}, depending on the oxide growth and annealing conditions. Large capture cross-section traps can also be found near to polysilicon/SiO_2 interfaces [198] and although their effect can be reduced by improved annealing conditions (they are observed to fill with very low applied biasses) this might be due to a reduction in the surface asperities of the polysilicon, rather than to changes in the traps themselves. These Coulombic traps have sometimes [114] been ascribed to a sodium centre, but the work of Ning et al [62] makes it much more probable that they are associated with the positive fixed interfacial charge. This is shown by the fact that the value of ΔV_T(sat) after electron injection is directly proportional to the initial value of Q_f.

Another Coulombic attractive trap, and one which does display a $E_{ox}^{-3/2}$ dependence of the capture cross-section, is the so-called 2.4 eV trap. This trap, which was studied in detail by Kapoor et al [66] is distributed uniformly throughout the oxide, with an effective density which is usually much smaller than that of the interfacial trap described above, and may even be completely absent in well-controlled processes. It has the unusual property of being optically accessible (2.4 eV below the oxide conduction band) and has been attributed to immobile sodium ions, especially in wet-grown oxides.

The many traps observed with cross-sections in the range 10^{-15}-10^{-18} cm^2 are usually electrically neutral, and are related to water in the oxidation atmosphere, and to radiation damage. Nicollian et al [193] studied these traps by exposing dry-grown oxides to water vapour before injecting electrons; and found that there was a linear relationship

Reference	Oxide Growth Conditions	Experimental Method	Capture Cross section, cm^2
Nicollian et al [193] 1971	p-type Si Dry O_2 1500-2500Å Water contaminated	Avalanche, etch-off	1.5×10^{-17}
Ning & Yu [61] 1974	p-type Si Dry O_2, 950°C 350-1500Å	Photo injection	(i) 3.3×10^{-13} (ii) 2.4×10^{-19}
Ning, Osburn & Yu [62] 1975	p-type (100) Dry O_2, 1000°C 500Å	Photo injection	3×10^{-13}
Arnett & Yun [195] 1975	Nitride 520Å Oxide 23Å	MNOS transistor characteristics	5×10^{-13}
Aitken & Young [196] 1976	p-type Si Dry O_2 500Å	Avalanche	(i) 10^{-14} (ii) 8×10^{-16} (iii) 4×10^{-18} (iv) 5×10^{-19}
DiMaria Aitken & Young [197] 1976	p-type Dry O_2 1000Å Sodium contaminated	Avalanche at 77K	(i) 2×10^{-15} (ii) 2×10^{-19} (iii) 5×10^{-20} plus others
Chaudhari [180] 1977	phosphorous doped n-channel process	B-T stressing of MOST	(i) 2.5×10^{-13}
Kapoor, Feigl & Butler [66] 1977	(100) n-type Wet O_2 1150°C 4.4 μm	Photo depopulation, etch-off	
Yun, Hickmott [198] 1977	(100) p-type Dry O_2 1000°C 350Å Poly Si Gate	Hysteresis of CV plots	
DiMaria, Ephrath & Young [199], 1979	(100) p-type Dry O_2 1000-1500Å Reactive ion etched	C-V plots, photo I-V	(i) 10^{-14} - 10^{-17} (ii) 10^{-14} - 10^{-18}
Ning [200] 1978	(100) p-type Dry O_2, HCl and Wet O_2, 500Å	Pulsed injection at 77K, Thermal depop	10^{-15}
Ning [201] 1978	(100) p-type 500Å, Dry O_2, 1000°C E-beam aluminium	Optical injection, V_T shift	(i) $> 10^{-13}$ (ii) 10^{-15}
Aitken Young & Pan [202], 1978	(100) p-type 440Å, Dry O_2, 100°C E-beam irradiated	Avalanche	(i) 1.2×10^{-14} (ii) 1.6×10^{-15} (iii) 2×10^{-16} (iv) 7×10^{-18} (v) 1×10^{-18}

Table 7.1 Summary of electron trap properties observed by various workers

Trap Density Note units	Trap Energy Level	Trap Spatial Distribution	Notes
Dependent on contamination		Near SiO_2 - air interface	
(i) 3.2×10^{16} cm^{-3} (ii) 2×10^{17} cm^{-3}		Uniform distribution assumed	
10^{11} cm^{-2}			Positive charge is trapping centre
6×10^{18} cm^{-3}			
(i) 10^{18} cm^{-2} (ii) 7×10^{10} cm^{-2} (iii) 2×10^{11} cm^{-2} (iv) 4×10^{11} cm^{-2}			Densities increase with x-ray stressing
Dependent on contamination			
(i) 1.5×10^{14} cm^{-3} (ii) $< 10^{17}$ cm^{-3}	2.4 eV below $E_C(SiO_2)$	(i) Uniform (ii) 100Å from both interfaces	(i) Optically accessible (ii) Total charge, most not optically accessible
	0.3 eV above E_F(poly Si)	30Å from poly Si-SiO_2 interface	
(i) $10^{9} - 10^{11}$ cm^{-2} (ii) $> 10^{13}$ cm^{-2}		(i) Uniform (ii) Near exposed oxide surface	(i) Oxide exposed to CF_4 plasma (ii) Oxide exposed to O_2 or A plasma
	0.3 eV below $E_C(SiO_2)$		Shallow traps at low temperatures
			(i) Positive charge centres (ii) Neutral traps
(i) 2×10^{10} cm^{-2} (ii) 4.4×10^{10} cm^{-2} (iii) 8.4×10^{10} cm^{-2} (iv) 2.6×10^{11} cm^{-2} (v) 4.7×10^{11} cm^{-2}			Cross sections of e-beam traps same at 77K and 298K

Table 7.1 (continued)

between the partial pressure of water vapour and the saturated threshold voltage shift. Because of their exposure method, the density of the traps in their experiments decreased exponentially from the SiO_2/air interface, but in normal oxides the density is constant in the bulk and rises sharply at both interfaces. According to their model, electro trapping occurs when an SiOH group captures on electron to become negatively charged SiO, with hydrogen being evolved from the oxide (as confirmed by radiotracer experiments with oxides contaminated with tritiated water). This electrochemic mechanism accounts for the fact that the traps are not significantly depopulated by 4 eV photons [192] although the charge can be thermally annealed with an activation energy of 0.35 eV, which is also the approximate activation energy for water vapour diffusion in SiO_2.

Exposure to radiation [196, 199, 201, 202, 217] or the implantation of ions [190, 191, 203, 204] gives rise to two classes of traps, one of which is Coulombic attractive and may be annealed out at ~400°C. The other radiation-induced traps have smaller (10^{-15}-10^{-18} cm^2) cross-sections, are electrically neutral and have spatial distributions which depend on the penetration of the radiation or charged species. These traps may only be annealed at temperatures above ~600°C, and are associated with bonding defects in the bulk oxide. The cross-sections have only weak temperatures and field dependencies, and the traps are energetically deep in the oxid band-gap.

The smallest traps, with cross-sections in the range 10^{-18}-10^{-21} cm^{-2}, are electrically neutral but have a repulsiv barrier, so they are only filled when very large fluxes of electrons are injected, and they may even be created by the large oxide currents required to detect them. DiMaria et al [197] have observed traps which they relate to sodium ions at the Si/SiO_2 interface, and which are only filled at 77K (although the trapped charge is stable when the sample is warmed to room temperature). Although the sodium ions are positively charged, it has been argued [154] that the small cross-sections could result from the screening effect produced by electrons in the silicon which are attracted to the ion.

Enhanced trapping at 77K has also been observed by other authors [186, 200]. In Ning's experiments [186], most of the trapped charge was thermally re-emitted at room temperature, t traps being located ~0.3 eV below the oxide conduction band. Young [200] observed that the enhanced charge was located ~90Å from the Si/SiO_2 interface. Approximately 10% of the additional charge captured at 77K remains trapped at room temperature. This effect could be due to an increase in capture cross-section with decreasing temperature. Alternatively, the electrons could be captured by traps which are too shallow to be observed at room temperature, with a subsequent deepening of the trap due lattice relaxation preventing any re-emission. The existence of these "deep" traps which are only observable at low temperatures does not seem [186] to depend on the oxide growth conditions (wet, dry, or HCl), suggesting that they are due to the intrinsic structure of the oxide.

7.4.3 Processing effects on trap densities

From a device reliability viewpoint, it is important to minimize the effective areal density of traps in the gate oxide, especially for the large capture cross-section traps which are most likely to capture electrons in service operation. Unfortunately, studies of the effects of processing parameters on trap densities reported in the literature do not normally distinguish between the various centres, but instead give a single figure representing the effective saturated trapping density (when all the traps are charged).

In general, oxides grown in dry oxygen have fewer traps than those grown in wet oxygen or steam [205], although for unannealed samples the increased number of fixed-charge related traps in the dry oxide may balance out the water-related traps in the wet oxide [155]. In-situ post-oxidation annealing at ~1000°C in nitrogen has been shown [206] to reduce the trapping observed in dry oxides at room temperature, but to increase it at 77K. This effect has not been explained, although as the cross-sections are in the 10^{-18} - 10^{-19} cm^2 range it is not particularly important for practical devices.

Post-metallization annealing at ~400°C in either nitrogen or forming gas is virtually essential to reduce the the density of the large (3×10^{-13} cm^2) cross-section, positive Coulombic attractive traps which exist at the Si/SiO_2 interface to allowable levels. These traps may result either from the oxide growth or from exposure to ionizing radiation. The smaller, neutral traps produced by radiation are more difficult to anneal as they result from atomic displacement, but their densities may be reduced [199, 202] by annealing in forming gas at temperatures above 500°C. Other traps, associated with implanted atoms [191] are unaffected by annealing at temperatures up to 1000°C.

In Chapter 6 it was shown that oxides grown in ambients containing HCl were very susceptible to hole trapping; their response to electron injection, however, is much less clear-cut. Iwamatsu and Tarui [155] found a decrease in the saturated trap density of HCl oxides compared to dry-O_2 ones, whilst Gdula [205] reported a slight increase. The HCl/H_2O oxides of Dorosti and Viswanathan [207] had much larger trap densitie than their dry-O_2 ones, largely due to the presence of Coulombic traps even after annealing, but these results might be due to the use of uv light for the photoinjection stage of the experiment.

Gate insulators other than thermal silicon dioxide have much larger trap densities. For example, Gdula and Li [208] observed the initial trapping efficiency of an Si_3N_4/SiO_2 insulator to be approximately 10^4 times larger than that of pure SiO_2, and this is confirmed by the results of Abbas and Dockerty [209], who modelled the threshold voltage shifts in MNOS transistors subjected to hot electron injection, and found a density of 5×10^{12} cm^{-2} Coulombic traps ($\sigma \sim 10^{-13}$ cm

These large cross-section traps are also responsible for the high trapping efficiency of chemical vapour deposited SiO_2 films [180], especially when they are doped with boron [205] or phosphorous [208]. They are again due to imperfections in the Si-O bonding, and their density can be reduced by high-temperature annealing. For devices with PSG/SiO_2 composite insulators, however, this annealing can cause diffusion of the phosphorous atoms towards the silicon, causing the trapping efficiency to rise again.

8

Conduction Problems and Dielectric Breakdown

One of the major attractions of MOS transistors is their very high input impedance and, although variations in growth conditions can give rise to large variations in the I-V characteristics of thermal silicon dioxide, even a poor oxide is not normally sufficiently leaky to cause any significant degradation of device performance. An exception to this is for non-volatile memory devices which rely on charge stored within an insulating layer as in this case both the writing of data and its long-term retention are critically dependent on the conduction properties of the insulator.

8.1 Thick Oxide MNOS Transistors and Double-Dielectric Instabilities

Silicon nitride (Si_3N_4) was originally suggested as an alternative to silicon dioxide as a gate insulator because of its higher dielectric constant, better masking of impurities, higher dielectric breakdown strength and because it presents a barrier to the drift of mobile ions. The properties of Si_3N_4 and methods for its deposition have been described by Deal et al [210]. MNS devices suffer from very large densities of surface states, a problem which is overcome by passivating the silicon surface with a layer of thermal SiO_2 before depositing the nitride, leading to the well-known MNOS structure.

MNOS devices may be divided into two classes, depending on the oxide thickness. Thick-oxide transistors (oxide thickness >100Å) have similar properties to ordinary MOS devices, but suffer from the "double dielectric" instability. In thin-oxide transistors (oxide thickness <50Å), this instability is utilised to produce a memory function; the modes of failure of this particular device are treated in the next section.

Thick-oxide MOSTs exhibit charge storage characteristics due to the trapping of charge at the oxide-nitride interface, originally described in 1969 by Frohman-Bentchkowsky and Lenzlinger [211], due to the different conductivities of the two insulators. The normal conduction mechanism in silicon dioxide is known [168] to be Fowler-Nordheim tunneling through the image-force lowered interface barrier, whereas conduction in the nitride is bulk-limited due to the field-enhanced thermal emission from shallow traps, that is, Frenkel-Poole conduction. An energy band diagram for conduction in an MNOS device with positive applied bias is shown in Figure 8.1. On initial application of the bias, the

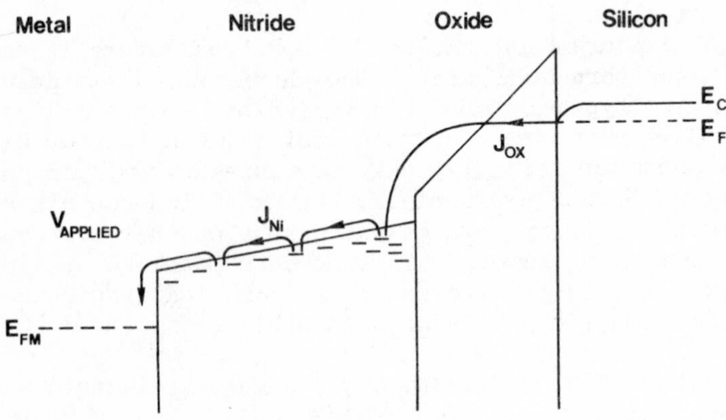

Figure 8.1 Band diagram for conduction through a thick-oxide MNOS device with positive applied bias.

fields in the oxide and nitride layers are determined by their relative thicknesses and their dielectric constants, giving rise (in general) to a current discontinuity. Charge thus accumulates at the interface until the internal fields reach values producing equal currents; this charge also, of course, produces a change in the flat-band voltage and hence constitutes another instability mechanism. The polarity of the stored charge depends on the relative thicknesses of the two layers but, for the more usual case of the nitride much thicker than the oxide, a positive applied voltage produces negative stored charge and vice versa, ie a positive hysteresis effect.

Theoretical values of the saturated threshold voltage shift and its time constant have been calculated by

Frohman-Bentchkowsky and Lenzlinger [211]. With sufficiently high fields (and for thin oxides), saturated values of ΔV_T can be in excess of 10V, but by careful adjustment of the insulator thicknesses and by limiting the operating conditions the shifts can be reduced to ~100 mV after 1000 hours. Although neither conduction mechanism has an Arrhenius-type temperature dependence, Chaudhari [104] has been able to model the threshold voltage shifts observed in the temperature range from 75°C to 165°C with the expression

$$\Delta V_T = \ln \left(\frac{t}{t_o}\right) \times K_o \exp\left(- \frac{\phi}{kT} + \beta E_n^{\frac{1}{2}}\right) \qquad (8.1)$$

where t_o, K_o and β are constants, E_n is the electric field in the nitride and ϕ is the activation energy which was 0.65 eV for the particular transistors used.

Several authors [104, 212, 213] have observed a further increase in the threshold shift after "saturation", and this is shown by the typical results in Figure 8.2. It has been suggested [104] that this additional shift is due to mobile ion motion as it is proportional to the square root of the biasing time, but the symmetry of the shifts in Figure 8.2 for both polarities indicates that this is probably not the case, unless the ions are initially distributed uniformly throughout the nitride. Instead, Kasprzak et al [213] have proposed that, during extended stressing, charge initially located at the oxide-nitride interface migrates into the bulk of the nitride and that the second saturation level is reached only when all the bulk traps are filled. Whatever the true origin of this additional charge, it results in an over-optimistic view of the stability of devices assessed using only short B-T stresses.

Further improvements to the stability of MNOS devices have been achieved [214] by "annealing" the nitride in oxygen or steam, thus producing a thin surface layer of SiO_2-Si_xN_y, which reduces the gate current.

Although the differential conduction instability is usually associated with MNOS devices, it is also present to some extent in all double dielectrics. For example, the instability of "slow polarization" of phosphosilicate glass deposited on thermal SiO_2 [118] is probably caused by this mechanism.

8.2 MNOS Charge Loss

When used as a memory device, the thin-oxide MNOS transistor must be able to trap both holes and electrons at the oxide-nitride interface, so that information may be both written and erased. Parameters which must be optimized include: the speed and ease of writing, long term retention of stored information,

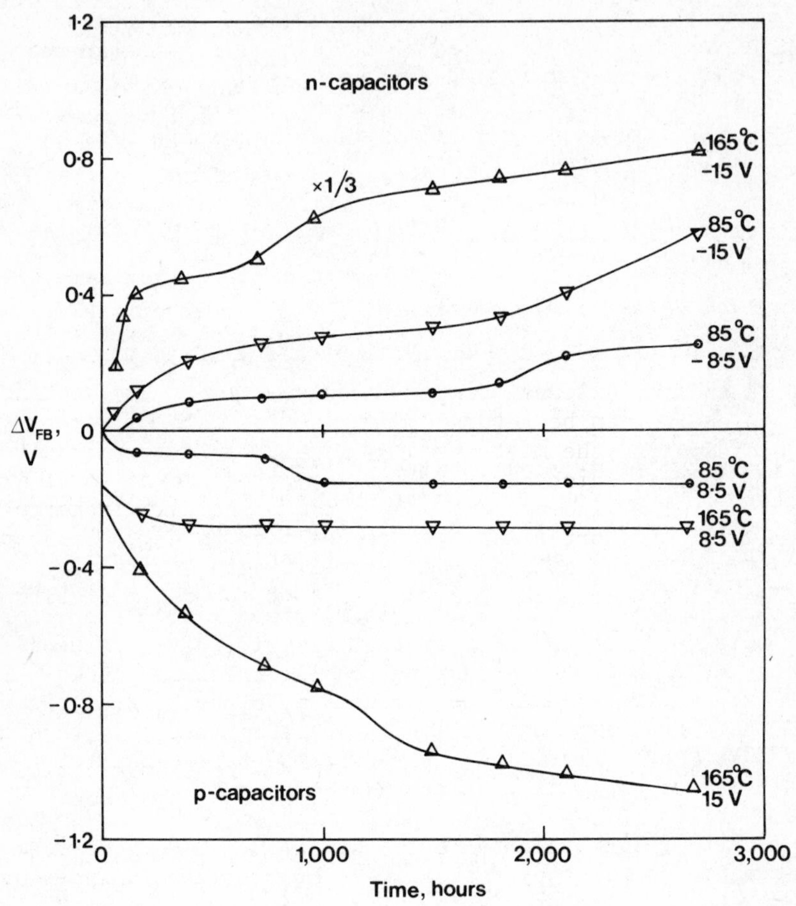

Figure 8.2 Flat-band voltage shifts for a thick (300Å) oxide MNOS device, showing an increase in the shift after the normal saturation. (After Kasprzak et al [212]).

ease of reading and the number of read cycles possible without refreshing the data, and the number of write/erase cycles poss before the written and the unwritten states become indistinguishable. In practice, these requirements are usuall achieved by a structure consisting of ~20Å of oxide and 400-800Å of nitride, and by writing and erasing with 1 ms pulses of ±20 to 30V.

The limitation on the number of write/erase cycles arises because of the build-up of interface states (which close the "memory window"), caused by the very high fields and the large oxide currents. This process has been reported by Woods and Tusca [94], and a method of investigating the energy distribution of the states without charging the dielectric is given by Arnold and Schauer [41a]. Even for a very good sample, and with carefully chosen reading voltages, it is difficult to exceed 10^{10} cycles [215], and this is only achieved by sacrificing the long-term storage time. For typical commercial MNOS memories [146], as few as 10^4 cycles can degrade the retention time from many tens of years to only 1000 hours.

After writing, the charge stored near the oxide-nitride interface will discharge under the influence of the built-in field and any external bias. There are several mechanisms by which charge can be returned from the insulator to the semiconductor, the most obvious of which is direct tunneling. This process, which has been studied by Lundkvist et al [216], is temperature independent and is the principal discharge mechanism at short times. As can be seen in Figure 8.3, the discharge begins at a characteristic time t_d, which is exponentially dependent on the oxide thickness, and continues linearly with logarithmic time. Arnett and Yun [195] found that, for a typical transistor with 20Å of oxide, the trapped charge is located in a distribution extending from the oxide-nitride interface into the bulk of the nitride, with traps in the first ~25Å of the nitride being completely filled at low levels of injected charge. Thus the logarithmic time decay is due to carriers tunneling directly from traps located deeper and deeper into the nitride.

Extrapolation of the direct tunneling mechanism leads some makers of commercial memories to make extravagant claims for charge retention but unfortunately at longer discharge times or at high temperatures a second and faster mechanism becomes important. Charge in the bulk of the nitride is detrapped by Frenkel-Poole emission, transported to the oxide-nitride interface by the built-in field, and then tunnels into the semiconductor. Lehovec and Fedotowsky [218] have modelled this mechanism by assuming a mono-energetic trap and have obtained reasonable agreement with the experimental behaviour. In the temperature range from 200-300°C, departure from the direct tunneling behaviour occurs after 1-10 seconds, the relatively small temperature dependence being attributed to the filling of shallow traps at low temperatures. The results of Lundkvist et al [219] which are shown in Figure 8.4 suggest that there is actually a rather broad energy

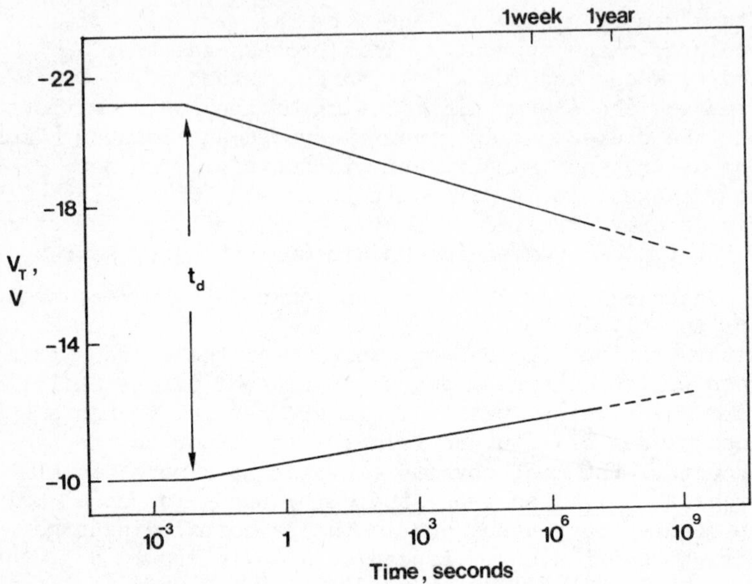

Figure 8.3 Discharge of a thin-oxide MNOS memory transistor by direct tunneling with no applied bias after writing with positive and negative pulses. (After Lundkvist et al [216]).

distribution of traps, with the exact distribution depending on the nitride deposition process.

The discharge process is accelerated by the application of a gate voltage of opposite polarity to that used during writing by two main mechanisms. Firstly, the increased electric field in the nitride reduces the effective depth of the traps by an amount $\beta\sqrt{E_N}$, where β is the Frenkel-Poole constant and E_N is the field. The second charge loss mechanism is due to a slow writing in the opposite direction to the original one. Maes and Van Overstraeten [220] have found that at low applied fields the injected currents (holes or electrons) are much larger than would be predicted on the basis of Fowler-Nordheim tunneling to the nitride bands through the oxide, and proposed that this is caused by tunneling into traps in the nitride, instead of directly into the conduction or valence band.

One further effect of applied fields on the data retentio of MNOS memory transistors is the rearrangement of charge within the nitride [221]. This rearrangement can also occur solely due to the built-in field, in which case the charge centroid moves towards the metal/nitride interface, reducing

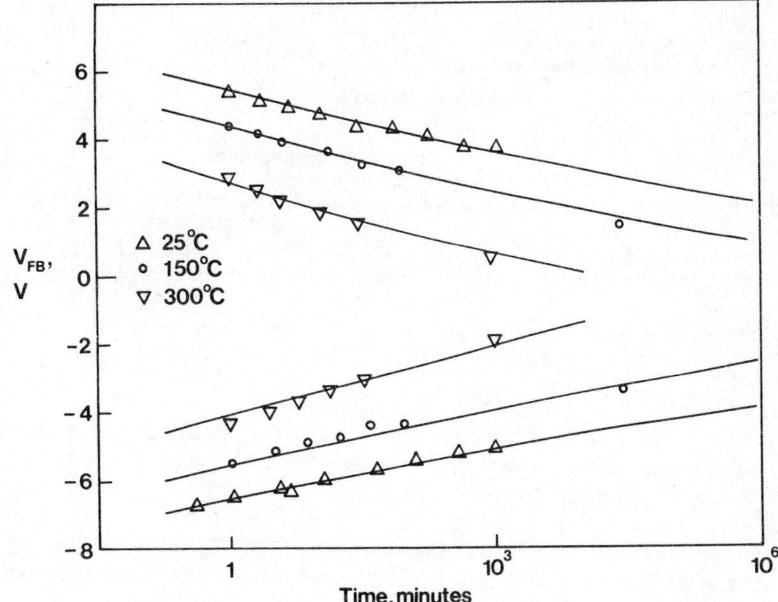

Figure 8.4 Thermal discharge of stored holes and electrons from a thin oxide (20Å) MNOS transistor, with no applied voltage. (After Lundkvist et al [219]).

the induced charge in the silicon, and also making it more difficult to erase the stored data.

8.3 Floating-gate Memory Devices

Floating-gate memory devices rely for their operation on the storage of electrons on a polysilicon MOST gate electrode which is completely isolated from the rest of the structure (the "floating-gate") by being surrounded by silicon dioxide. The earliest devices [222] were based on p-channel transistors and achieved electron injection to the floating-gate during the writing of data by avalanching the drain junction, as was described in Chapter 7. Later n-channel devices, illustrated in Figure 8.5, employ the channel hot-electron mechanism, the transistor being turned "on" by means of a voltage applied to a second "select" gate stacked on top of the isolated floating-gate. Erasure of the stored information in both types of device is achieved by exposure to uv illumination, which causes the trapped electrons to be re-emitted to the silicon.

The voltages applied to the memory transistor to sense the presence of stored charge during reading are similar to those used during writing, but are smaller in magnitude.

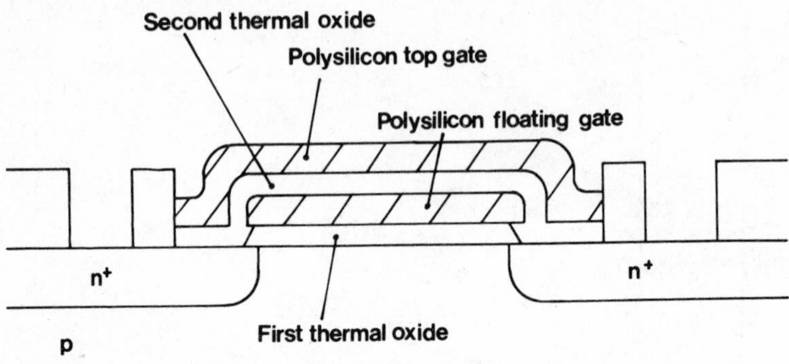

Figure 8.5 Diagrammatic cross-section of a double-polysilicon floating-gate memory transistor. The metallization and passivation layers have been omitted for clarity.

This can give rise to an unintentional writing of previously unwritten cells, causing a loss of data after extended periods of reading. The results of Jeppson and Svensson [223] for p-channel junction avalanche devices are shown in Figure 8.6, where it can be seen that the mechanism has a large voltage dependence and is accelerated by operation at low temperatures. The temperature dependence has the empirical form [223]

$$t = t_o \exp(0.025T). \tag{8.2}$$

Similar results for n-channel devices have been reported by Watt [146] and are shown in Figure 8.7. Extrapolation of these results to the normal reading voltage of 5V produces a time of approximately 3 years continuous reading for a 1V change in the floating gate voltage. This is an adequate period for most purposes, but in practice allowance must be made for variations in the channel length between different transistors, which can result in large changes in the electron injection rate.

Long-term losses of stored charge in floating-gate memori can occur due to leakage of the injected electrons through the second-level thin oxide to the select gate or through the first-level oxide to the substrate, source or drain, according to the applied bias. (It is also conceivable that slow hole trapping, resulting from the built-in field of the negative charge, could compensate for the stored charge, but there have been no unequivocal observations of this mechanism). It is often observed that oxides grown on poly-crystalline silicc

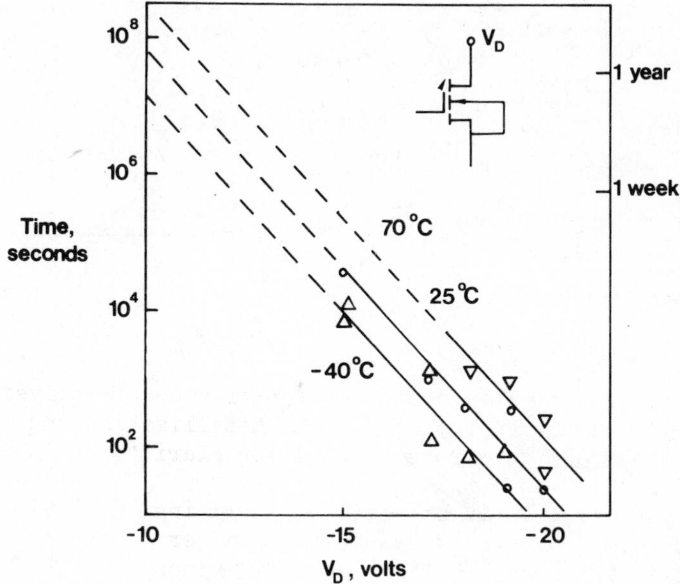

Figure 8.6 Unintentional writing of a p-channel memory transistor, showing the time required for a 2V change in the floating-gate voltage. (After Jeppson and Svensson [223]).

conduct heavily at lower applied voltages than similar oxides grown on single-crystal silicon [224-226]. This is especially true when electron injection occurs from the silicon, as is shown in Figure 8.8. It has been found [226] that the enhanced conduction exhibits the same Fowler-Nordheim characteristics observed for oxides grown on single-crystal silicon [168] and that the interface barrier height (measured with an internal photoemission technique) also has a similar value. However, whereas for oxides grown on single crystal silicon the interfacial field is simply given by dividing the applied voltage by the oxide thickness, for poly-oxides there are localized regions of the interface with fields much higher than this average value, due to surface asperities on the surface of the polysilicon. Anderson and Kerr [225] have observed these asperities by scanning electron microscopy, and typically found 5×10^9 asperities per cm^2 with an average height of 750Å in films where the oxide was grown at 1000°C, but no asperities higher than ~100Å for oxides grown at 1150°C. These results correlate well with the conduction observed in the two films, although even for the higher

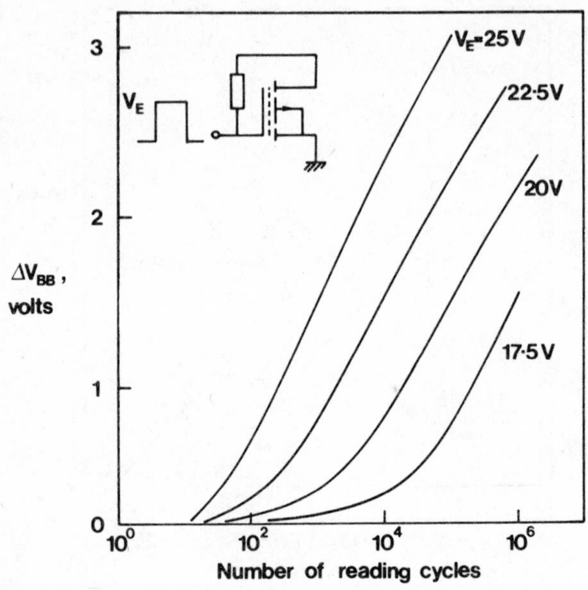

Figure 8.7 Unintentional writing of an n-channel floating-gate memory transistor by reading with excessive read voltages. Read pulse length = 1 ms. (After Watt [146]).

temperature oxide the conduction is larger than for the single crystal case. Indeed, slightly enhanced injection has also been observed [227] into oxides underlying polysilicon layers, perhaps due to higher fields associated with the grain boundaries.

Conduction through poly-oxide is complicated by electron trapping in the oxide near the regions of high injection, which causes a power-law time dependence of the current. The I-V characteristics can be determined without the effect of charge trapping by using a pulsed technique [228] to minimize the charge flow from the floating gate. The effect of charge trapping on a memory transistor is to increase the data retention time, and to make it virtually independent of the initial level of stored charge [227]. It has been suggested [229] that electrically-alterable memories could be based on the removal of stored charge from the floating-gate by applying large positive voltages to the select gate, but charge trapping in the oxide prevents repeated erasure.

Because Fowler-Nordheim tunneling does not have an exponential temperature dependence, care must be taken when

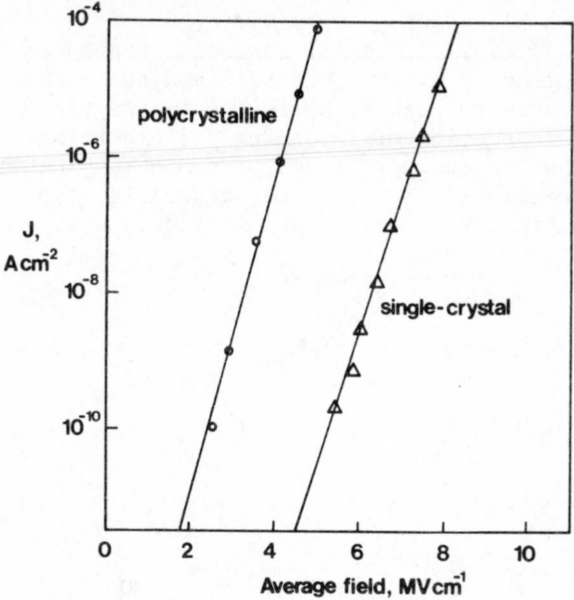

Figure 8.8 Conduction through silicon-SiO_2-aluminium structure with 500Å of oxide and with the aluminium biased positively. (After DiMaria and Kerr [226]).

making predictions of long-term data retention based on high-temperature accelerated tests. Many authors assume an Arrhenius relationship, based on experiments over a limited temperature range, but Alexander [230] found that the apparent activation energy was 1 eV in the 200-300°C range and 0.5 eV between 85 and 125°C, and Davis [227] reported only a factor of ten difference in the time required to lose 10% of the stored charge between 77K and room temperature.

Degradation of the retention time of random cells in large memory arrays has been observed [146] after repetitive reading. The current in these substandard cells was almost directly proportional to the applied voltage, and is thought to occur at defects along the edges of the floating-gate, where the second-level thin oxide is grown in contact with the first-level oxide. A similar case of anomolous leakage at the edges of self-aligned single-level polysilicon gate devices has been reported [231], in this case with a Schottky-type relationship (log I:$\sqrt{V}$), and with the values of the current and the apparent barrier height showing wide variations with the processing conditions. The exact cause of these excessive leakages is not known.

8.4 Dielectric Breakdown

Dielectric breakdown is very different from the other instabilities already discussed, not only because it is an intrinsic property of silicon dioxide (and all other insulators), but also because it is a catastrophic failure mode, with no indication of impending failure before final breakdown. The breakdown voltages of nominally identical MOS capacitors taken from the same wafer generally have distributions similar to that shown in Figure 8.9. Most

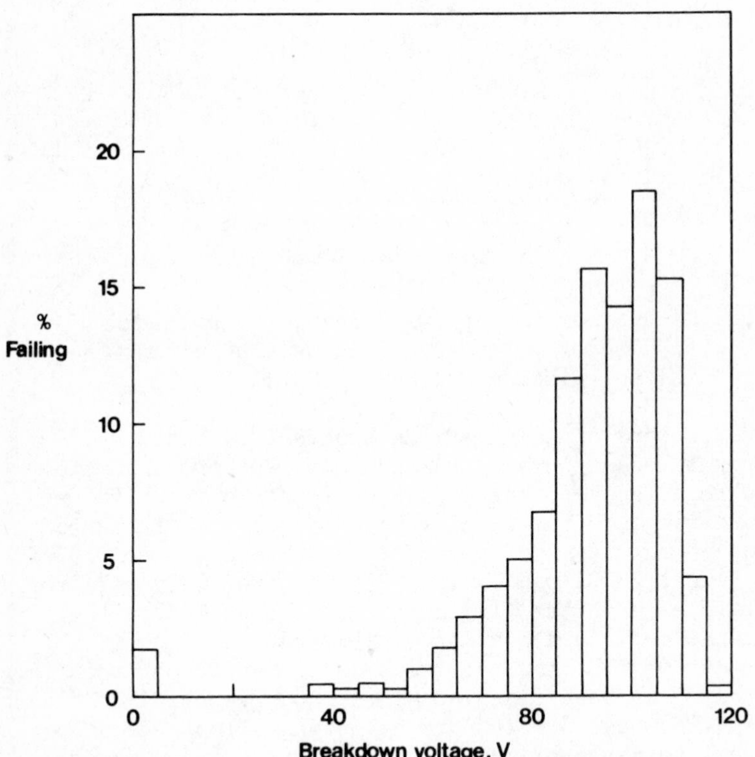

Figure 8.9 Typical breakdown voltage histogram for MOS capacitors with an oxide thickness of 1350Å and an area of 0.08 mm^2.

capacitors breakdown at voltages well in excess of those applied in practical integrated circuits, so that only the small proportion with very low breakdown voltages (which are usually thought to result from "defects" in the insulator) are of direct concern to reliability studies.

A bibliography of dielectric breakdown in thin films has been given by Agarwal [231].

8.4.1 Intrinsic Breakdown

Although the intrinsic (short-term) dielectric strength of SiO_2 does not in itself present a reliability hazard, it does set an upper limit to the gate voltage that can be applied, and the mechanism of intrinsic breakdown is of interest as it may provide an insight into the causes of defect-induced breakdown.

It is widely believed [159, 235, 236] that breakdown in wide bandgap insulators such as SiO_2 is initiated by impact ionization, possibly aided by the space-charge field due to the virtually immobile [237] holes produced by the avalanche process. Although the electron velocity in SiO_2 "saturates" for fields between 8×10^5 and 5×10^6 V/cm due to lattice scattering, at higher fields the velocity rises sharply again due to field-induced modifications of the scattering [325]. At $(7-8) \times 10^6$ V/cm velocity runaway occurs, and carrier multiplication gives rise to large localized current densities which ultimately cause vapourization of the oxide. The final breakdown site is very small (~1 μm in diameter), and the time-to-failure, which results from the build-up of the high-energy tail of the electron distribution and the statistical nature of the avalanche process, is very short (<<1 second).

For very thin (<100Å) oxides, breakdown cannot be caused by impact ionization because the maximum voltage which can be supported by the oxide is insufficient to allow an electron travelling from one electrode to the other without suffering a lattice collision to gain the ionization energy (~12 eV [235]). Instead, Harari [172] has proposed that breakdown occurs in thin films when the internal field exceeds the Si-O bond strength, at about 3×10^7 V/cm. When the average field impressed on the oxide is less than this value, electron trapping near the injecting interface causes a gradual increase in the field in the bulk until failure occurs.

8.4.2 Defect-induced Breakdown

Capacitors which breakdown at voltages very much lower than the intrinsic dielectric strength (and which produce the low-voltage tail of Figure 8.9) contain oxide defects. The distribution of defects across a wafer is the subject of some controversy but, for high quality substrates and processing, it has been shown [238] that they are randomly distributed and that the probability of finding a defect in a capacitor of a known area is given by the Poisson distribution. (Defects can also occur in clusters due, for example, to scratches in the substrate). Defects not only produce a reduction in the yield of integrated circuits, but can also produce a reliability hazard if they degrade further during life. Two classes of long-term failure mechanisms may be envisaged. Firstly, the defect itself may degrade (physically or structurally) so that it can no longer withstand the same electric field. Secondly, movement or trapping of charge may occur within the oxide until the total field exceeds the breakdown voltage of the defective region.

The principal defect cited in the literature [107, 124, 239-241] is that caused by sodium ion contamination. For purposely contaminated samples, the oxide current with a positive applied field reaches a maximum as the sheet of sodium reaches a critical distance (~35Å) from the Si/SiO_2 interface, and Raider [239] has been able to correlate this increased electron tunneling current with the incidence of dielectric breakdown. In order to cause failure, this mechanism requires a minimum concentration of sodium which is much higher than is normally seen in uncontaminated samples so for practical devices a second mechanism involving clustering of the ions at local interface irregularities [125, 126] has to be evoked. There are additional reasons to believe that dielectric failure of integrated circuits in service conditions may not be due to mechanisms other than sodium contamination, notably the insensitivity of breakdown to the polarity of the field [242], and the activation energy for failure [248], which is much smaller than that for sodium motion.

Another possible degradation mechanism is devitrification of the oxide, a process which Meek and Braun [24] have observed to be accelerated by the presence of sodium, especial in steam-grown oxides. Other mechanisms which have been proposed include particulate inclusions [244, 245], hot-spots due to cathodic protruberences [246], and the localized build-up of stationary holes [236]. None of these mechanisms has yet been developed into a model capable of explaining service failures.

8.4.3 Acceleration of Dielectric Breakdown

There is very little agreement in the literature on the nature of the time dependence of the breakdown probability, an its acceleration by high temperatures and high electric fields This is due to the wide variability of the times-to-failure of nominally identical samples, which is related to the defects having a range of severities, and to the presence of at least two failure mechanisms.

Early investigators [247], who used very high stress leve (both high temperature and high voltage) to obtain sufficient numbers of failures for analysis from small batch sizes, reported that failures were only observed with positive bias applied to the top (metal) electrode. In these conditions, the cumulative breakdown probability increases linearly with logarithmic time, allowing a maximum time for 100% failures to be defined. The dependence of this time on the field and voltage for a 200Å thick oxide is shown in Figure 8.10. Li et al [107] have developed a theory, based on the field

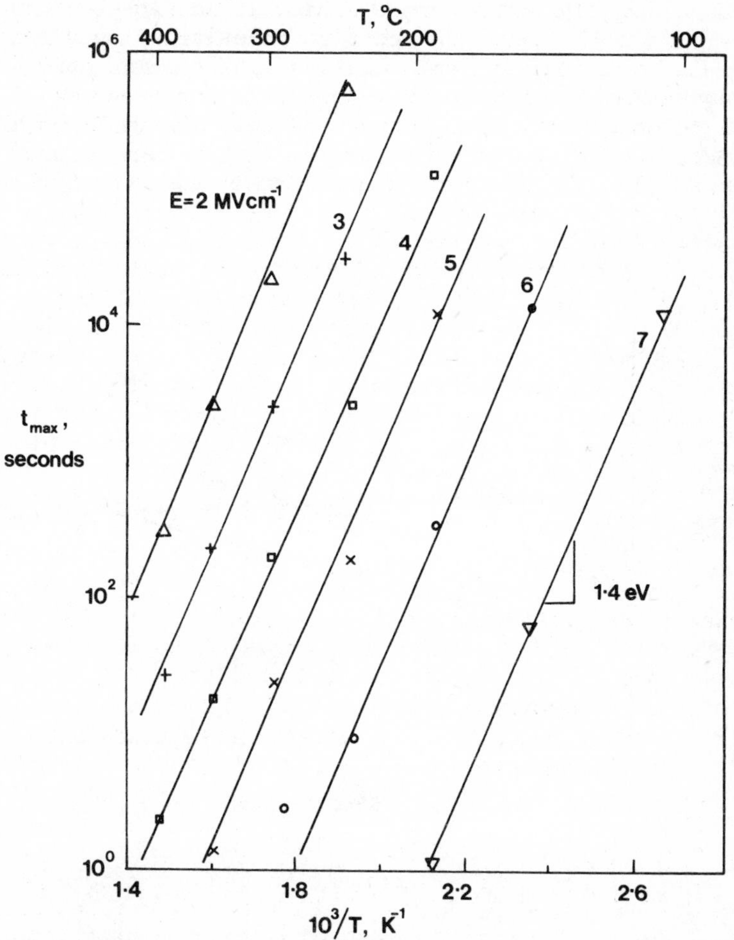

Figure 8.10 Temperature and field dependence of the limiting time-to-breakdown for Al-200Å SiO_2-Si capacitors with positive bias. (After Osburn and Chou [247]).

induced emission of trapped ions, which predicts that 100% failures will be observed in a finite time and appears to give a good fit to experimental results obtained with either contaminated samples or where the substrate has purposely been damaged by ion implantation. An alternative explanation of these high-stress failures, originally proposed by Osburn and Chou [247], is that they result from barrier lowering produced by the build-up of states near the interface. These states, which are produced by the high oxide currents flowing at the high stress levels, could be associated with the anomolous positive charge observed by Young et al [186].

By using very large sample sizes, it has now been shown [242, 243] that failures do occur at relatively low stress levels for both positive and negative applied bias, and it is thought that the mechanism responsible for these failures is the same as that causing integrated circuit failures in service. As can be seen in Figure 8.11, the cumulative failure probability exhibits a log-normal type distribution,

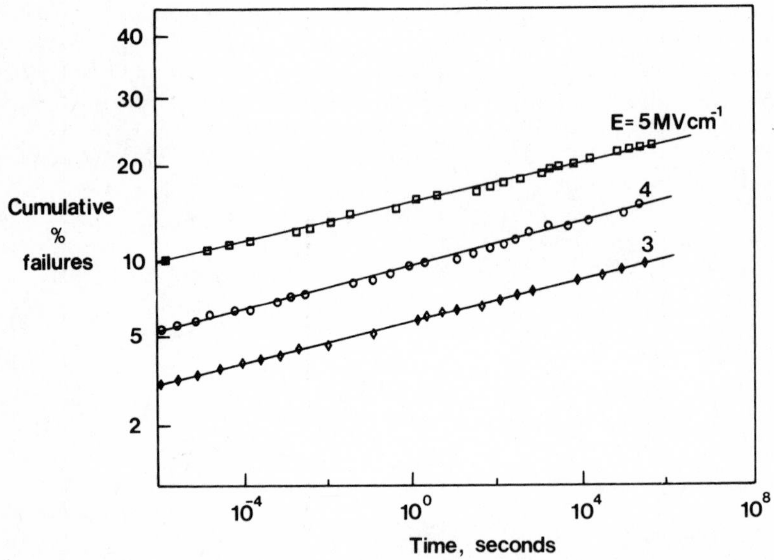

Figure 8.11 Room-temperature time-dependent dielectric breakdown of 1100Å thick oxides with negative applied bias. (After Crook [243]).

and there is a very large dispersion (shallow slope), so that it is never possible to observe 100% failure. There are two empirical approaches to modelling these results. Crook [243] has used the standard log-normal failure density function and introduced additional factors to account for the temperature and field acceleration by assuming an Arrhenius temperature dependence (with an experimentally determined activation energy of 0.35 eV), and by using an electric field acceleration factor A_F given by

$$A_F = \exp\left(\frac{E_s - E_o}{0.062 \times 10^6 \text{ V/cm}}\right) \tag{8.3}$$

where E_S is the stress field and E_o is the normal operating field. Anolick and Nelson [242] have adopted the different approach of relating the time-to-breakdown of a given percentage of the sample to the voltage required for the same percentage of identical devices to fail during a fast voltage ramp measurement. The shape of the time-to-breakdown distribution thus depends on the shape of the low-voltage tail of the breakdown voltage histogram (such as Figure 8.9), although in practice at low failure levels it is indistinguishable from a log-normal distribution. This model has the disadvantage that the "activation energy" obtained depends on the oxide area of the samples, which can have no physical basis, but it is based on the not unreasonable assumption that the most severe defects also degrade the fastest, accounting for the large dispersion. Because of the large acceleration factors involved, it will not be possible to select the most appropriate model for extrapolating accelerated life test data to service conditions until the actual mechanism for polarity-independent breakdown is discovered.

References

The references quoted in the main text are organised into various topics in the table below

Topic	Reference numbers
Reviews/Text books/ Bibliographies	5-9,15,16,21,136,139,231
Basic theory	10-13,17,47,54,55,181
Preparation techniques	20,22,82,119,120,155,214
Oxide and interface structure	21,23-37,74,92,120,160
Measurement techniques	10,12,35,39-51,54-60,63,66,69,72,73, 75,109,148,190,194
Surface states	10,22,36,38-51,53,55,57,60,67-92,94, 122,135,152,161,198,215
Mobile ions	24,56-59,78,95-133,138,141,145,147
Dipolar polarization	117,118,134-138,210
Hole trapping	36,52,63-65,78,90a,131,137,139-144, 149-157,161
Electron injection and trapping	61,62,66,154,155,158,162-209,213,216, 217
Double dielectrics	41a,81,90a,94,104,118,136-138,163, 195,210-221
Conduction mechanisms	113,115,148,149,157-159,162,168,169, 171,210,211,218,222-229,231,234-237
Dielectric breakdown	24,92,107,124,125,144,159,172,231, 234-248
Radiation effects	19,36,38,61,64,67,70,93,160,171,196, 199,201,202,204,217

NUMERICAL LISTING

1. LILIENFELD J E: US Patent No 1745175, (1930)

2. HEIL O: British Patent No 439137, (1935)

3. SHOCKLEY W and PEARSON G L: Modulation of conductance of thin films of semiconductor by surface charges. Phys Rev 74, pp 232, (1948)

4. KAHNG D and ATTALA M M: Silicon-silicon dioxide field induced surface devices. IRE Solid State Dev Res Conf (1960)

5. SCHLEGEL E S: A bibliography of metal-insulator-semiconductor studies. IEEE Tran Electron Dev ED-14, No 11, pp 728-49, (November 1967)

6. AGAJANIAN A H: A bibliography on silicon dioxide films. Solid State Technol, pp 36-48, (January 1977)

7. PANTELIDES S T (Ed): The Physics of SiO_2 and its interfaces. Pub Pergamon, (1978)

8. SZE S M: Physics of semiconductor devices. Pub Wiley New York, (1969)

9. MANY, GOLDSTEIN and GROVER: Semiconductor surfaces. Pub Wiley New York, (1965)

10. NICOLLIAN E H and GOETZBERGER A: The Si-SiO_2 interface - electrical properties as determined by the MIS conductan technique. Bell Syst Tech J, 46, pp 1055-1133, (1967)

11. GOETZBERGER A: Ideal MOS curves for silicon. Bell Syst Tech J, 45, pp 1097-1122, (1966)

12. GOETZBERGER A and NICOLLIAN E H: Temperature dependence inversion layer frequency response in silicon. Bell Sys Tech J, pp 513-22, (March 1967)

13. DEAL B E and SNOW E H: Barrier energies in metal-silicon dioxide-silicon structures. J Phys Chem Solids, 27, pp 1873-9, (1966)

14. GROVE A S: Physics and technology of semiconductor devic Pub Wiley (1967)

15. CRAWFORD R H: MOSFET in circuit design. Pub McGraw-Hill, (1967)

16. CARR W N and MIZE J P: MOS/LSI design and application. Pub McGraw-Hill, (1972)

17. RICHMAN P: Theoretical threshold voltages for MOS field effect transistors. Solid-state Electron, 11, pp 869-76, (1968)

18. TROUTMAN R R: Subthreshold design considerations for insulated gate field effect transistors. IEEE J Solid State Ccts SC-9, No 2, pp 55-60, (April 1974)

19. Numerous papers presented at the IEEE Conf on nuclear and space radiation effects, pub in IEEE Trans Nuc Sci NS-23, No 6, (December 1976)

20. GAIND A K, ACKERMANN G K, LUCARINI U J and BRATTER R L: Preparation and properties of SiO_2 films from $SiH_4CO_2H_2$. J Electrochem Soc, 123, No 1, pp 111-7, (January 1976)

21. REVERZ A G: Non-crystalline silicon dioxide films on silicon-a review. J Non-Cryst Solids, 11, pp 309-30, (1973).

22. RAZOUK R R and DEAL B E: Dependence of interface state density on silicon thermal oxidation process variables. J Electrochem Soc 126, No 9, pp 1573-81, (September 1979)

23. CLARKE R A, TAPPING R L, HOPPER M A and YOUNG L: An ESCA study of the oxide at the Si-SiO_2 interface. J Electrochem Soc 122, No 10, pp 1347-50, (October 1975)

24. MEEK R L and BROWN R H: Devitrification of steam-grown silicon dioxide films. J Electrochem Soc 119, No 11, pp 1538-44, (November 1972)

25. ALESSANDRINI E I and CAMPBELL D R: Catalysed Crystallization in SiO_2 thin films. J Electrochem Soc 121, No 8, pp 1115-8, (August 1974)

26. JOHANNESSEN J S, SPICER W E and STRAUSSER Y E: An auger analysis of the Si-SiO_2 system. J Appl Phys 47, No 7, pp 3028-37, (July 1976)

27. BLANC J, BUIOCCHI C J, ABRAHAMS N S and HAM W E: The Si/SiO_2 interface examined by cross-sectional transmissio electron microscopy. Appl Phys Lett 30, No 2, pp 120-2, (January 1977)

28. KRIVANEK O L, SHENG T T and TSUI D C: A high-resolution electron microscopy study of the $Si-SiO_2$ inteface. Appl Phys Lett 37, No 7, pp 437-9, (April 1978)

29. HELMS C R, SPICER W E and JOHNSON N M: New studies of the $Si-SiO_2$ interface using auger sputter profiling. Solid State Commun 25, No 9, pp 673-6, (March 1978)

30. HELMS C R, STRAUSSER Y E and SPICER W E: Observation of an intermediate chemical state of silicon in the Si/SiO_2 interface by auger sputter profiling. Appl Phys Lett 33, No 8, pp 767-9, (October 1978)

31. HELMS C R, JOHNSON N M, SCHWARZ S A and SPICER W E: Studies of the effect of oxidation time and temperature o the $Si-SiO_2$ interface using auger sputter profiling. J Appl Phys 50, No 11, pp 7007-14, (November 1979)

32. OFFERMANN P: Thickness evaluation of Si/SiO_2 interfaces by He-backscattering experiments. J Appl Phys 48, No 5, pp 1890-4, (May 1977)

33. CHEUNG N W, FELDMAN L C, SILVERMAN P J and STENSGAARD: Studies of the $Si-SiO_2$ interface by MeV ion channeling. Appl Phys Lett 35, No 11, pp 859-61, (December 1979)

34. IRENE E A: Some observations of defects in amorphous SiO_2 films. in The Physics of SiO_2 and its Interfaces, Ed Pantelides, Pub Pergamon, pp 205-9

35. REVESZ A G: On SiOH and SiH groups in SiO_2 films on silicon. J Electrochem Soc 124, No 11, pp 1811-3, (November 1977)

36. SVENSSON C M: The defect structure of the $Si-SiO_2$ interfa a model based on trivalent silicon and its hydrogen "compounds". in The Physics of SiO_2 and its Interface, Ed Pantelides, Pub Pergamon, pp 328-32, (1978)

37. WHITE C T and NGAI K L: Metastabilities at the $Si-SiO_x$ interface. in The Physics of SiO_2 and its Interfaces, Ed Pantelides, Pub Pergamon, pp 412-6, (1978)

38. McGARRITY J M, WINOKUR P S, BOESCH H E and McLEAN F B: Interface states resulting from a hole flux incident on the Si/SiO_2 interface. in The Physics of SiO_2 and its Interfaces, Ed Pantelides Pub Pergamon, pp 428-33, (1978)

39. TERMAN L M: An investigation of surface states at a silicon/silicon dioxide interface employing metal-oxide-silicon diodes. Solid state Electron 5, pp 285-299, (1962)

40. BERGLUND C N: Surface states at steam-grown silicon-silicon dioxide interfaces. IEEE Tran Electron Dev ED-13, No 10, pp 701-5, (October 1966)

41. KUHN M: A quasi-static technique of MOS C-V and surface state measurements. Solid State Electron 13, pp 873-85, (1970)

41a. ARNOLD E and SCHAUER H: Measurements of interface state density in MNOS structures. Appl Phys Lett 32, No 5 pp 33-5, (March 1978)

42. BROWN D M and GRAY P V: Si-SiO_2 fast interface state measurements. J Electrochem Soc 115, No 7, pp 760-6, (July 1968)

43. GRAY P V and BROWN D M: Density of SiO_2-Si interface states. Appl Phys Lett 8, No 2, pp 31-3, (January 1966)

44. FONTELLA GONCALVES N and CHARRY E: Influence of Φ_{ms}(T) in the Determination of the Si-SiO_2 surface states. IEEE Trans Electron Dev 26, No 9, pp 1377-8, (September 1979)

45. AMELIO G F: A new method for measuring interface state densities in MIS devices. Surface Science 29, pp 125-44, (1972)

46. WANG K L and EVWARAYE A O: Determination of interface and bulk-trap states of IGFETs using deep level transient spectroscopy. J Appl Phys 47, No 10, pp 4574-7, (October 1976)

47. ZAMANI N and MASERJIAN J: Lateral nonuniformities (LNU) of oxide and interface state charge. in The Physics of SiO_2 and its Interfaces, Ed Pantelides, Pub Pergamon, pp 443-8, (1978)

48. WERNER C, BERNT H and EDER A: Inhomogeneities of surface potential in the thermally grown Si-SiO_2 interface. J Appl Phys 50, No 11, pp 7015-19, (November 1979)

49. BRUGLER J S and JESPERS P G: Charge pumping in MOS devices. IEEE Trans Electron Dev ED-16, No 3, pp 297-302, (March 1969)

50. ELLIOT A B M: The use of charge-pumping currents to measure surface-state densities in MOS transistors. Solid-state Electron 19, pp 241-7, (1976)

51. OWCZAREK A and KOLODZIEJSKI J F: Charge pumping effect and its application in the MOS transistors investigations Electron Technol 10, No 1, pp 55-70, (1977)

52. REYNOLDS F H: Room temperature threshold voltage instabilities in an MOS integrated circuit type MP104B. Post Office Research Report No 409, (1973)

53. SEQUIN C and BALDINGER E: Interface states in Metal-Oxide-Semiconductor Field Effect Transistors. Solid-state Electron 13, pp 1527-40, (1970)

54. VAN OVERSTRAETEN R J, DECLERCK G J and BROUX G L: The influence of surface potential fluctuations on the operation of the MOS transistor in weak inversion. IEEE Trans Electron Dev, ED-20, No 12, pp 1154-8, (December 1973)

55. VAN OVERSTRAETEN R J, DECLERCK G J and MULS P A: Theory of the MOS transistor in weak inversion - new method to determine the number of surface states. IEEE Trans Electron Dev, ED-22, No 5, pp 282-8, (May 1975)

56. CHOU N J: Application of triangular voltage-sweep method mobile charge studies in MOS structures. J Electrochem Soc 118, No 4, pp 601-9, (April 1971)

57. PHILLIPS W E, KOYAMA R Y and BUEHLER M G: Suppression of measurement interferences from interface states and mobile ions in thermally stimulated current measurements in an MOS capacitor. J Electrochem Soc 126, No 11, pp 1979-81, (November 1979)

58. HICKMOTT T W: Thermally stimulated ionic conductivity of sodium in thermal SiO_2. J Appl Phys 46, No 6, pp 2583-98, (June 1975)

58a. BOUDRY M R and STAGG J P: The kinetic behaviour of mobile ions in the Al-SiO_2-Si system. J Appl Phys 50, No 2, pp 942-50, (February 1979)

59. NAUTA P K and HILLEN M W: Investigation of mobile ions in MOS structures using the TSIC method. J Appl Phys 49, No 5, pp 2862-5, (May 1978)

60. YAMASHITA K, IWAMOTO M and HINO T: A method for studying interface states in MIS structures by thermally stimulated surface potential. J Appl Phys 49, No 5, pp 2866-75, (May 1978)

61. NING T H and YU H N: Optically induced injection of hot electrons into SiO_2. J Appl Phys 45, No 12, pp 5373-8, (December 1974)

62. NING T H, OSBURN C M and YU H N: Electron-trapping at positively charged centres in SiO_2. Appl Phys Lett 26, No 5, pp 248-50, (March 1975)

63. NING T H: Capture cross-section and trap concentration of holes in silicon dioxide. J Appl Phys 47, No 3, pp 1079-81, (March 1976)

64. POWELL R J: Hole photocurrents and electron tunnel injection induced by trapped holes in SiO_2 films. J Appl Phys 46, No 10, pp 4557-63, (October 1975)

65. WOODS M H and WILLIAMS R: Hole traps in silicon dioxide. J Appl Phys 47, No 3, pp 1082-9, (March 1976)

66. KAPOOR V J, FEIGL F J and BUTLER S R: Energy and spatial distribution of an electron trapping centre in the MOS insulator. J Appl Phys 48, No 2, pp 739-49, (February 1977)

67. HUGHES G W: Interface-state effects in irradiated MOS structures. J Appl Phys 48, No 12, pp 5357-9, (December 1977)

68. ZIEGLER K: Distinction between donor and acceptor character of surface states in the Si-SiO_2 interface. Appl Phys Lett 32, No 4, pp 249-51, (February 1978)

69. SCHULZ M and JOHNSON N M: Transient capacitance measurements of hole emission from interface states in MOS structures. Appl Phys Lett 31, No 9, pp 622-5, (November 1977)

70. FAHRNER W and GOETZBERGER A: Properties of a single level surface state induced by Be implantation into Si-SiO_2 interfaces. J Appl Phys 44, No 2, pp 727-7, (February 1973)

71. DEULING H, KLAUSMAN E and GOETZBERGER A: Interface states in Si-SiO_2 interfaces. Solid-state Electron 15, pp 559-71, (1972)

72. MORITA M, TSUBOUCHI K and MIKOSHIBA: Measurement of interface-state parameters near the band edge at the Si/SiO_2 interface by the conductance method. Appl Phys Lett 33, No 8, pp 745-7, (October 1978)

73. MULS P A, DECLERCK G H and VAN OVERSTRAETEN R J: Influence of interface charge inhomogeneities on the measurement of surface state densities by means of the MOS ac conductance technique. Solid-state Electron 20, pp 911-22, (1977)

74. LAUGHLIN R B, JOANNOPOULOS J D and CHADI D J: Electronic states of Si-SiO_2 interfaces. in The Physics of SiO_2 and its Interfaces, Ed Pantelides, Pub Pergamon, pp 321-7, (1978)

75. LAM Y W: Interface studies of the MIS structure by surface photovoltage measurements. Electronics Lett 6, No 6, pp 153-4, (March 1970)

76. ARNOLD E: Surface charges and surface potential in silicon inversion layers. IEEE Trans Electron Dev ED-15, No 12, pp 1003-8, (December 1968)

77. BREWS J R: Surface potential fluctuations generated by interface charge inhomogeneities in MOS devices. J Appl Phys 43, No 5, pp 2306-13, (May 1972)

78. DEAL B E, SKLAR M, GROVE A S and SNOW E H: Characteristics of the surface-state charge (Q_{ss}) of thermally oxidised silicon. J Electrochem Soc 114, No 3, pp 266-73, (March 1967)

79. WANG K L: A study of quench-in defects and interface states of MOS structures. Tech Digest IEEE Electron Devices Mtg, pp 154-8, (1977)

80. LANE C H: Stress at the Si-SiO_2 interface and its relationship to interface states. IEEE Trans Electron Dev ED-15, No 12, pp 998-1003, (December 1968)

81. DEAL B E, McKENNA E L and CASTRO P L: Characteristics of fast surface states associated with SiO_2 and Si_3N_4-SiO_2 structures. J Electrochem Soc 116, No 7, pp 997-1005, (July 1969)

82. CASTRO P L and DEAL B E: Low temperature reduction of fast surface states associated with thermally oxidised silicon. J Electrochem Soc 118, No 2, pp 280-6, (Bebruary 1971)

83. SCHLEGEL E S: Effects of aluminium linewidth on the annealing of fast states in MOS structures. IEEE Trans Electron Dev ED-19, No 6, pp 839-40, (June 1972)

84. HICKMOTT T W: Annealing of surface states in polycrystalline silicon gate capacitors. J Appl Phys 48, No 2, pp 723-33, (February 1977)

85. SEVERI M and SONCINI G: Surface state density at the (hydrogen chloride) oxide-silicon interface. Electronic Lett 8, No 16, pp 402-4, (August 1972)

86. GOETZBERGER A, LOPEZ A D and STRAIN R J: On the formation of surface states during stress ageing of thermal Si-SiO_2 interfaces. J Electrochem Soc 120, No 1, pp 90-6, (January 1973)

87. SAMINADAYAR K and PFISTER J C: Evolution of surface-states density of Si/wet thermal SiO_2 interface during bias-temperature treatment. Solid-state Electron 20, pp 891-6, (1977)

88. SEMUSHKINA N A and SEMUSHKIN G B: Energy spectrum of the surface-states in the Si-SiO_2 system. Sov Phys Solid state 15, No 1, pp 1-5, (July 1973)

89. KASSABOU J D and ILIEVA M N: Effects of bias temperature treatments on the metal-SiO_2-Si system. Comptes Rendus de l'Academi Bulgare des Sciences 25, No 7, pp 897-90, (1972)

90. KOBAYASHI I, NAKAHARA M and ATSUMI M: Study of the Si-SiO_2 interface state with the negative bias-heat treatment approach. Proc IEEE 61, pp 249-50, (February 1973).

90a. JEPPSON K O and SVENSSON C M: Negative bias stress of MOS devices at high electric fields and degradation of MNOS devices. J Appl Phys 48, No 5, pp 2004-14, (May 1977)

91. SHIONO N, NAKAJIMA O, HASHIMOTO C and MURAMOTO S: New threshold voltage instability in n-MOSFETs. Private Communication

92. JORGENSEN P J: Electrolysis of SiO_2 on silicon. J Chem Phys 48, No 4, pp 1594-8, (August 1968)

93. HUGHES H L: A survey of radiation induced perturbations in metal-insulator-semiconductor structures. Proc 9th Ann IEEE Rel Phys Symp, pp 33-9, (1971)

94. WOODS M H and TUSKA J W: Degradation of MNOS memory transistor characteristics and failure mechanism model. Proc 10th Ann IEEE Rel Phys Symp, pp 120-5, (1970)

95. SNOW E H, GROVE A S, DEAL B E and SAH C T: Ion transport phenomena in insulating films. J Appl Phys 36, No 5, pp 1664-73, (May 1965)

96. HOFSTEIN S R: Proton and sodium transport in SiO_2 films. IEEE Trans Electron Dev ED-14, No 11, pp 749-59, (November 1967)

97. HINO T and YAMASHITA K: Neutralization of mobile ions in the SiO_2 film of MOS structures. J Appl Phys 50, No 7, pp 4879-82, (July 1979)

98. STAGG J P and BOUNDRY M R: The neutralization of Na^+ ions in HCl grown SiO_2. Rev Physique Appl 13, pp 841-3, (December 1978)

99. GUTHRIE J W, WELLS U A and DERBENWICH G F: Potassium contamination in MOS fabrication. Extd. Abs. Fall Mtg. Electrochem Soc, pp 349-51, (October 1976)

100. YON E, KO W and DRUPER A: Sodium distribution in thermal oxide on silicon by radiochemical and MOS analysis. IEEE Trans Electron Dev ED-13, pp 276-80, (February 1966)

101. CARLSON H, FULLER C, OSBORNE J and BROWN G A: Stability of etched oxides. Extd Abs Fall Mtg Electrochem Soc, pp 177, (1966)

102. ELDRIDGE J M and KERR D R: Sodium ion drift through phosphosilicate glass - SiO_2 films. J Electrochem Soc 118, No 6, pp 986-91, (June 1971)

.03. GRUNTHANER F J and MASERJIAN J: Sodium ions at defect sites at Si-SiO_2 interfaces as determined by X-ray photoelectron spectroscopy. Proc 13th Ann IEEE Rel Phys Symp, pp 15-25, (1975)

.04. CHAUDHARI P K: Threshold voltage degradation of MNOS FET devices. J Electrochem Soc 125, No 10, pp 1657-60, (October 1978)

.05. RAI B P and SRIVASTAVA R S: Mobile ion instability in SiO_2 films on silicon. Int J Electronics 46, No 4, pp 381-92, (1979)

06. HILLEN M W: Dynamic behaviour of mobile ions in SiO_2 layers. in The Physics of SiO_2 and its interfaces, Ed Pantelides, Pub Pergamon, pp 179-83, (1978)

.07. LI S P, PRUSSIN S and MASERJIAN J: Model for MOS field-time dependent breakdown. Proc IEEE 16th Ann Rel Phys Symp, pp 132-136, (1978)

08. STAGG J P: Drift mobilities of Na^+ and K^+ ions in SiO_2 films. Appl Phys Lett 31, No 8, pp 532-3, October 1977

08a HILLEN M W, GREEUW G and VERWEY: On the mobility of potassium ions in SiO_2. J Appl Phys 50, No 7, pp 4834-7, (July 1979)

09. WOODS M H and WILLIAMS R: Injection and removal of ionic charge at room temperature through the interface of air with SiO_2. J Appl Phys 44, No 12, pp 5506-10, (December 1973)

10. GREENWOOD C J: Control of field threshold voltage by caesium placement within the field dielectric. Eigth European Solid State Device Research Conference, pp 391-3, (1978)

11. STAGG J P and BOUDRY M R: Interfacial diffusion of Na^+ ions at the Si-SiO_2 interface, and Na^+ neutralization in the presence of chlorine. Insulating Films on Semiconductors, Inst Phys Conf Ser No 50, pp 75-80, (1979)

12. TANGENA A G, DE ROOIJ N F and MIDDELHOEK J: Sensitivity of MOS structures for contamination with H^+, Na^+ and K^+ ions. J Appl Phys 49, No 11, pp 5576-83, (November 1978)

113. TANGENA A G, MIDDELHOEK J and DE ROOIJ N F: Influence of positive ions on the current-voltage characteristics of MOS structures. J Appl Phys 49, No 5, pp 2876-9, (May 1978)

114. MAYO S and EVANS W H: Development of sodium contaminatio in semiconductor oxidation atmospheres at 1000°C. J Electrochem Soc 124, No 5, pp 780-5, (May 1977)

115. DIMARIA D J: Room-temperature conductivity and location of mobile sodium ions in the thermal silicon dioxide layer of a metal-silicon dioxide-silicon structure. J Appl Phys 48, No 12, pp 5149-51, (December 1977)

116. KERR D R, LOGAN J S, BURKHARDT P J and PLISKIN W A: Stabilization of SiO_2 passivation layers with P_2O_5. IBM J Res Dev 8, No 4, pp 376-84, (September 1964)

117. SNOW E H and DEAL B E: Polarisation phenomena and other properties of phosphosilicate glass films on silicon. J Electrochem. Soc 113, No 3, pp 263-9, (March 1966)

118. CHAUDHARI P K, MICHAUD R A and QUINN R M: Stability of MOSFET devices with phosphorus-doped oxide as gate dielectric. J Electrochem Soc 124, No 12, pp 1897-1900, (December 1977)

119. KRIEGLER R J, CHENG Y C and COTTON D R: The effect of HCl and Cl_2 on the thermal oxidation of silicon. J Electrochem Soc 119, No 3, pp 388-92, (March 1972)

120. MONKOWSKI J, TRESSLER R E and STACH J: The structure and composition of silicon oxides grown in HCl/O_2 ambients. J Electrochem Soc 125, No 11, pp 1867-73, (November 1978

121. ROHATGI A, BUTLER S R and FEIGL F J: Na neutralization characteristics of HCl oxides in MOS structures. Extd Abs Electrochem Soc Spring Mtg, pp 219-20, (1977)

122. TOPICH J A; Compensation of mobile ion movement in SiO_2 by ion implantation. Appl Phys Lett 33, No 11, pp 967-9 (December 1978)

123. SILVERSMITH D J: Non-uniform lateral ionic impurity distributions at Si-SiO_2 interfaces. J Electrochem Soc 119, No 11, pp 1589-93, (November 1972)

124. DISTEFANO T H: Dielectric breakdown induced by sodium in MOS structures. J Appl Phys 44, No 1, pp 527-8, (January 1973)

125. WILLIAMS R and WOODS M H: Laser-scanning photoemission measurements of the silicon-silicon dioxide interface. J Appl Phys 43, No 10, pp 4142-7, (October 1972)

126. WILLIAMS R and WOODS M H: Image forces and the behaviour of mobile positive ions in silicon dioxide. Appl Phys Lett 22, No 9, pp 458-9, (May 1973)

127. NEMETH-SALLAY M, SZABO R, SZEP I C and TUTTO P: Charge motion in ether-treated silicon MOS structures. Presented at 4th Symposium on Solid State Device Technology, Munich (1979)

128. NAKAYAMA H, SHINDO M and ISHIKAWA T: Organic compounds inducing the room temperature instabilities of p-channel silicon gate MOS transistors. J Electrochem Soc 126, pp 1301-3, (July 1979)

129. NAKAYAMA H, OSADA Y and SHINDO M: Room temperature instabilities of p-channel silicon gate MOS transistors. J Electrochem Soc 125, No 8, pp 1302-6, (August 1978)

130. NAKAYAMA H and SHINDO M: Mechanism of the room temperature instability of p-channel silicon gate transistor. Extd Abs Electrochem Soc Spring Mtg 248-9, (1977)

131. DUNN P J: Long-term threshold voltage instabilities in p-channel silicon-gate MOS transistors. British Post Office Research Report, (1980)

132. WATT A S M and ELLIOT A B M: Non-uniform threshold voltage instabilities in p-channel silicon gate MOS transistors. Electronics Lett 11, No 23, pp 559, (1975)

132a. SCHLEGEL E S and SCHNABLE G L: A negative ion type instability in MOS devices. J Electrochem Soc 119, No 2, pp 165-8, (February 1972)

133. KAPLAN L H and LOWE M E: Phosphosilicate glass stabilization of MOS structures. J Electrochem Soc 118, No 10, pp 1649-53, (October 1971)

134. ELDRIDGE J M, LAIBOWITZ R R and BALK P: Polarization of thin phosphosilicate glass films in MGOS structures. J Appl Phys 40, No 3, pp 1922-30, (March 1969)

135. REYNOLDS F H: Room temperature threshold-voltage instabilities in an MOS integrated circuit. Proc 12th Ann IEEE Rel Phys Symp, pp 170-6, (1974)

136. WOODS M H: Instabilities in double dielectric structures Proc 12th Ann IEEE Rel Phys Symp, pp 259-66 (1974)

137. LAMPI E E and LABUDA E F: A reliability study of insulated gate field effect transistors with an Al_2O_3-SiO_2 gate structure. Proc 10th Ann IEEE Rel Phys Symp, pp 112-9 (1972)

138. GRADINGER A D and ROSENZWEIG W: Polarization and charge motion in metal-Al_2O_3-SiO_2-Si structures. J Electrochem Soc 121, No 5, pp 700-5, (May 1974)

139. NICOLLIAN E H: Interface instabilities. Proc 12th Ann IEEE Rel Phys Symp, pp 267-72 (1974)

140. SINHA A K and SMITH T E: Kinetics of the slow-trapping instability of the Si/SiO_2 interface. J Electrochem Soc 125, No 5, pp 743-5 (May 1978)

141. HOFSTEIN S R: Stabilization of MOS Devices. Solid-State Electron 10, pp 657-70, (1967)

142. RAI B P and SRIVASTAVA R S: Negative bias instability in SiO_2 films on silicon. J Phys D. 11, pp 2139-46 (1978)

143. BREED D J: Non-ionic room temperature instabilities in MOS devices. Solid-State electron. 17, pp 1229-43, (1974)

144. SASAKI N: Change of Si-SiO_2 interface charge by B-T treatment. Jpn J Appl Phys 12, No 9, pp 1458-9, (1973)

145. REYNOLDS F H: The response of the threshold voltages of the transistors in simple MOS circuits to tests at elevated temperatures. Proc 9th Ann IEEE Rel Phys Symp, pp 46-56, (1971)

146. WATT A S M: Private communication

147. DERBENWICK G F: Mobile ions in SiO_2: Potassium. J Appl Phys 48, No 3, pp 1127-30, (March 1977)

148. WALDEN R H: A method for the determination of high-field conduction laws in the presence of charge trapping. J Appl Phys 43, No 3, pp 1178-86, (March 1978)

149. WEINBERG Z A, JOHNSON W C and LAMPERT M A: High-field transport in SiO_2 on silicon induced by corona charging of the unmetallized surface. J Appl Phys 47, No 1, pp 248-55, (January 1976)

150. DIMARIA D J, WEINBERG Z A and AITKEN J M: Location of positive charges in SiO_2 films on Si generated by VUV photons, x-rays, and high field stressing. J Appl Phys 48, No 3, pp 898-906, (March 1977)

151. HOLMES-SIEDLE A G and GROOMBRIDGE I: Positive charge traps in silicon dioxide films; a comparison of population by x-rays and band-gap light. Thin Solid Films 27, pp 165-70, (1975)

152. WEINBERG Z A and RUBLOFF G W: Exciton transport in SiO_2. in The Physics of SiO_2 and its Interfaces, Ed Pantelides, Pub Pergamon, pp 24-8, (1978)

153. WEINBERG Z A, YOUNG D R, DIMARIA D J and RUBLOFF G W: Exciton or hydrogen diffusion in SiO_2?. J Appl Phys, 50, No 9, pp 5757-60, (September 1979)

154. DIMARIA D J: The properties of electron and hole traps in thermal silicon dioxide layers grown on silicon. in The Physics of SiO_2 and its Interfaces, Ed Pantelides Pub Pergamon, pp 160-78, (1978)

155. IWAMATSU S and TARUI Y: A method for reducing the hole and electron trapping densities in thermal SiO_2 films. J Electrochem Soc, 126, No 6, pp 1078-80, (June 1979)

156. DE KEERSMAECKER R F and DIMARIA D J: Hole trapping in ion-implanted SiO_2. Presented at 4th Symposium on Solid State Device Technology, Munich (1979)

157. CURTIS O L and SROUR J R: The multiple-trapping model and hole transport in SiO_2. J Appl Phys 48, No 9, pp 3189-3828, (September 1977)

158. SOLOMON P M and AITKEN J M: Current and C-V instabiliti in SiO_2 at high fields. Appl Phys Lett. 31, No 3, pp 215-6 (August 1977)

159. DISTEFANO T H and SHATZKES M: Dielectric instability an breakdown in SiO_2 thin films. J Vac Sci Technol 13, No 1, pp 40-4, (January/February 1976)

160. LELL E, KREIDL N J and HENSLER J R: Radiation effects i quartz, silica and glasses. Progress in Ceramic Sci 4, pp 1-94, (1966)

161. HESS D W: Effect of chlorine on the negative bias instability in MOS structures. J Electrochem Soc 124, No 5, pp 740-3, (May 1977)

162. HUGHES R C: High field electronic properties of SiO_2. Solid-State Electron 21, pp 251-8, (January 1978)

163. NING T H, OSBURN C M and YU H N: Threshold instabilitie in IGFETs due to emission of leakage electrons from silicon substrate into silicon dioxide. Appl Phys Lett 29, No 3, pp 198-200, (August 1976)

164. DAVIS J R: Degradation behaviour of n-channel MOSFETs operated at 77K. To be published.

165. COTTRELL P E, TROUTMAN R R and NING T H: Hot-electron emission in n-channel IGFETs. IEEE J Solid-State Ccts, SC-14, No 2, pp 442-455, (April 1979)

166. ABBAS S A and DAVIDSON E E: Reliability implications of hot electron generation and parasitic bipolar action in an IGFET device. Proc 14th Ann IEEE Rel Phys Symp, pp 18-22, (1976)

167. FURUYAMA T, OHUCHI K and KOHYAMA S: An electrical mechanism for holding time degradation in dynamic MOS RAMS. IEEE Trans Electron Dev ED-26, No 11, pp 1684-90 (November 1979)

168. LENZLINGER M and SNOW E H: Fowler-Nordheim tunneling in thermally grown SiO_2. J Appl Phys 40, No 1, pp 278-83, (January 1969)

169. WEINBERG Z A: Tunneling of electrons from Si into therm grown SiO_2. Solid-state Electron 20, pp 11-18, (1977)

70. SOLOMON P M: High-field electron trapping in SiO_2. J Appl Phys 48, No 9, pp 3843-9, (September 1977)

71. HARARI E: Conduction and trapping of electrons in highly stressed ultrathin films of thermal SiO_2. Appl Phys Lett, 30, No 11, pp 601-3, (June 1977)

72. HARARI E: Dielectric breakdown in electrically stressed thin films of thermal SiO_2. J Appl Phys 49, No 4, pp 2478-89, (April 1978)

73. HICKMOTT T W: Temperature dependence of the relaxation of injected charge at the polycrystalline-silicon-SiO_2 interface. J Appl Phys 49, pp 3392-6, (June 1978)

74. NING T H, OSBURN C M and YU H N: Emission probability of hot electrons from silicon into silicon dioxide. J Appl Phys 48, No 1, pp 286-93, (January 1977)

75. ABBAS S A and DOCKERTY R C: Hot electron induced degradation of n-channel IGFETs. Proc 14th Ann IEEE Rel Phys Symp, pp 38-41, (1976)

76. ABBAS S A and DOCKERTY R C: Hot-carrier instability in IGFETs. Appl Phys Lett 27, No 3, pp 147-8, (August 1975)

77. EUZENT B: Hot electron injection efficiency in IGFET structures. Proc 15th Ann IEEE Rel Phys Symp, pp 1-4, (1977)

78. DAVIS J R: Temperature and voltage acceleration of hot-electron trapping in n-channel MOSFETs. Unpublished.

79. NING T H, COOK P W , DENNARD R H, OSBURN C M, SCHUSTER S E and YU H N: 1 μm MOSFET VLSI Technology: Part IV - Hot-electron design constraints. IEEE J Solid-state Ccts, SC-14, No 2, pp 268-75, (April 1979)

80. CHAUDHARI P K: Leakage-induced hot carrier instability in phosphorous doped SiO_2 gate IGFET devices. Proc 15th Ann IEEE Rel Phys Symp, pp 5-9, (1977)

81. DUNN P J and MELLOR P J T: The degradation of MOS transistors resulting from junction avalanche breakdown. Microelectronics and Reliability 11, pp 367-76, (1972)

182. COE D J: Changes in effective-channel length due to hot-electron trapping in short channel MOSTs. S-S and Electron Dev 2, No 2, pp 57-61, (March 1978)

183. PHILLIPS A, O'BRIEN R R and JOY R C: IGFET hot electron emission model. Tech Digest IEEE Electron Dev Conf, pp 38-42, (1975)

184. BULUCEA C: Avalanche injection into the oxide in silicon gate-controlled devices - I Theory. Solid-state Electron, 18, pp 363-74, (1975)

185. NICOLLIAN E H, GOETZBERGER A and BERGLUND C N: Avalanche injection currents and charging phenomena in thermal SiO_2. Appl Phys Lett, 15, No 6, pp 174-7, (September 19

186. YOUNG D R, IRENE E A, DIMARIA D J, DEKEERSMAECKER R F and MASSOUD H Z: Electron trapping in SiO_2 at 295K and 77K J Appl Phys 50, No 10, pp 6366-72, (October 1979)

187. VERWEY J F, KRAMER R P and de MAAGT B J: Mean free path of hot electrons at the surface of boron-doped silicon. J Appl Phys, 46, No 6, pp 2612-2619

188. YOUNG D R: Electron current injected into SiO_2 from p-ty Si depletion regions. J Appl Phys, 47, No 5, pp 2098-21 (May 1976)

189. BULUCEA C: Avalanche injection into the oxide in silicon gate-controlled devices - I Theory. Solid-State Electro 18, pp 363-74, (1975)

190. DIMARIA D J, YOUNG D R, DEKEERSMAECKER R F, HUNTER W R and SERRANO C M: Centroid location of implanted ions in the SiO_2 of MOS structures using the photo I-V technique. J Appl Phys 49, No 11, pp 5441-4, (November 1978)

191. DIMARIA D J, YOUNG D R, HUNTER W R and SERRANO C M: Location of trapped charge in aluminium-implanted SiO_2. IBM J Res Develop 22, No 3, pp 298-293, (May 1978)

192. POWELL R J and BERGLUND C N: Photoinjection studies of charge distributions in oxides of MOS structures. J Appl Phys 42, No 11, pp 4390-7, (October 1971)

193. NICOLLIAN E H, BERGLUND C N, SCHMIDT P F and ANDREWS J M: Electrochemical charging of thermal SiO_2 films by injected electron currents. J Appl Phys 42, No 13, pp 5654-64, (December 1971)

194. POWELL R J: On the determination of charge centroids in insulators by photoinjection or photodepopulation. Appl Phys Lett 31, No 4, pp 290-1, (August 1977)

195. ARNETT P C and YUN B H: Silicon nitride trap properties as revealed by charge-centroid measurements on MNOS devices. Appl Phys Lett 26, No 3, pp 94-6, (February 1975).

196. AITKEN J M and YOUNG D R: Electron trapping by radiation-induced charge in MOS devices. J Appl Phys 47, No 3, pp 1196-8, (March 1976)

197. DIMARIA D J, AITKEN J M and YOUNG D R: Capture of electrons into Na^+-related trapping sites in the SiO_2 layers of MOS structures at 77K. J Appl Phys 47, No 6, pp 2740-3, (June 1976)

198. YUN B H and HICKMOTT T W: Charge injection from polycrystalline silicon into SiO_2 at low fields. J Appl Phys 48, No 2, pp 718-22, (February 1977)

199. DIMARIA D J, EPHRATH L M and YOUNG D R: Radiation damage in silicon dioxide films exposed to reactive ion etching. J Appl Phys 50, No 6, pp 4015-21, (June 1979)

200. NING T H: Thermal re-emission of trapped electrons in SiO_2. J Appl. Phys 49, No 12, pp 5997-6003, (December 1978)

201. NING T H: Electron trapping in SiO_2 due to electron-beam deposition of aluminium. J Appl Phys 49, No 7, pp 4077-82, (July 1978)

202. AITKEN J M, YOUNG D R and PAN K: Electron trapping in electron-beam irradiated SiO_2. J Appl Phys 49, No 6, pp 3386-3391, (June 1978)

203. YOUNG D R, DIMARIA D J and BOJARCRZUK N A: Electron-trapping characteristics of W in SiO_2. J Appl Phys 48, No 8, pp 3425-7, (August 1977)

204. DEKEERSMAECKER R F, DIMARIA D J and PANTELIDES S T: Photodepopulation of electrons trapped in SiO_2 on sites related to As and P implantation. in The Physics of SiO_2 and its Interfaces, Ed Pantelides, Pub Pergamon, pp 185-93 (1978)

205. GDULA R A: The effects of processing on hot electron trapping in SiO_2. J Electrochem Soc 123, No 1, pp 42-47 (January 1976)

206. YOUNG D R: Electron trapping in SiO_2. Inst Phys Conf Ser No 50, pp 28-39, (1980)

207. DOROSTI J and VISWANATHAN C R: Photo-injection studies of traps in HCl/H_2O oxides. in The Physics of SiO_2 and its Interfaces, Ed Pantelides, Pub Pergamon, pp 184-7, (1978)

208. GDULA R A and LI P C: Hot-electron trapping in CVD PSG films. J Electrochem Soc, 124, No 12, pp 1927-30, (December 1977)

209. ABBAS S A and DOCKERTY R C: N-channel IGFET design limitations due to hot electron trapping. Tech Digest IEEE Electron Dev Conf, pp 35-8, (1975)

210. DEAL B E, FLEMING P J and CASTRO P L: Electrical properties of vapour deposited silicon nitride and silicon oxide films on silicon. J Electrochem Soc 115, No 3, pp 300-7, (March 1968)

211. FROHMAN-BENTCHKOWSKY D and LENZLINGER M: Charge transport in Metal-Nitride-Oxide (MNOS) structures, J Appl Phys 40, No 8, pp 3307-19, (July 1969)

212. KASPRZAK L A, GAIND A K and HORNUNG A : Pseudo-stable MNOS structure. Extd. Abs Fall Mtg Electrochem Soc, pp 313-5, (October 1975)

213. KASPRZAK L A, GAIND A K and HORNUNG A: Pseudostable MNOS structures. J Electrochem Soc 124, No 10, pp 1631- (October 1977)

214. DOCKERTY R C, BARILE C A, NAGARAJAN A and ZALAR S M: Improved V_t stability of SNOS FETS by oxygen annealing. Proc 11th Ann IEEE Rel Phys Symp, pp 159-62, (1973)

215. WHITE M H, DZIWIANSKI J W and PECKERAR M C: Endurance of thin-oxide nonvolatile MNOS memory transistors. Proc IEEE Electron Dev Conf, pp 177-80, (December 1976)

216. LUNDKUIST L, LUNDSTROM C and SVENSSON C: Discharge of MNOS structures. Solid-State Electron 16, pp 811-23, (1973).

217. AITKEN J M: 1 μm MOSFET VLSI technology: Part VIII - Radiation effects. IEEE Trans Electron Dev, ED-26, No 4, pp 372-9, (April 1979)

218. LEHOVIC K and FEDOTOWSKY A: Charge retention of MNOS devices limited by Frenkel-Poole detrapping. Appl Phys Lett 32, No 5, pp 335-8, (March 1978)

219. LUNDKVIST L, SVENSSON C and HANSSON B: Discharge of MNOS structures at elevated temperatures. Solid-State Electron 19, pp 221-7, (1976)

220. MAES H E and VAN OVERSTRAETEN R J: Low field transient behaviour of MNOS devices. J Appl Phys 47, No 2, pp 664-66, (February 1976)

221. MAES H E and VAN OVERSTRAETEN R J: Memory loss in MNOS capacitors. J Appl Phys 47, No 2, pp 667-71, (February 1976)

222. FROHMAN-BENTCHKOWSKY D: A fully-decoded 2048-bit electrically programmable MOS ROM. Proc IEEE Solid-State Ccts Conf pp 80-2 (1971)

223. JEPPSON K O and SVENSSON C M: Unintentional writing of a FAMOS memory device during reading. Solid-State Electron 19, pp 455-7 (1976)

224. ABBAS S A, BARILE C A and DOCKERTY R C: Conductivity of poly-oxide. Extd Abs Electrochem Soc Fall Mtg 76, No 2, pp 842-3, (1976)

225. ANDERSON R M and KERR D R: Evidence for surface asperity mechanism of conductivity in oxide grown on poly-crystalline silicon. J Appl Phys 48, No 11, pp 4834-6, (November 1977)

226. DIMARIA D J and KERR D R: Interface effects and high conductivity in oxides grown from polycrystalline silicon. Appl Phys Lett, 27, No 9, pp 505-7, (November 1975)

227. DAVIS J R: Discharge mechanisms in floating-gate EPROM cells. Electronics Lett, 15, No 1, pp 20-21, (January 1979)

228. HU C, SHUM Y, KLEIN T and LUCERO E: Current-field characteristics of oxides grown from polycrystalline silicon. Appl Phys Lett 35, No 2, pp 189-91, (July 1979)

229. HZUKA H, MASUOKA F, SATO T and ISHIKAWA M: Electrically alterable avalanche-injection-type MOS read only memory with stacked gate structure. IEEE Trans Electron Dev ED-23, No 4, pp 379-87, (April 1976)

230. ALEXANDER R M: Accelerated testing in FAMOS devices - 8K EPROM. Proc IEEE 16th Ann Rel Phys Symp, pp 229-32, (1978)

231. HAM W E and EATON S S: Anomalous electrical gate conduction in self-aligned MOS strcutures. Tech Digest IEEE Electron Dev Mtg, pp 323-6, (1976)

232. AGARWAL V K: Breakdown conduction in thin dielectric films: A bibliographic survey. Thin Solid Films, 24, pp 55-70 (1974)

234. DIMARIA D J and KERR D R: Interface effects and high conductivity in oxides grown from polycrystalline silicon. Appl Phys Lett 27, No 9, pp 505-7, (November 1975)

235. FERRY D K: Electron transport and breakdown in SiO_2. J Appl Phys, 50, No 3, pp 1422-1427, (March 1979)

236. DISTEFANO T H and SHATZKES M: Dielectric instability and breakdown in wide bandgap insulators. J Vac Sci Technol, 12, No 1, pp 37-46, (January 1975)

237. HUGHES R C: Hole mobility and transport in thin SiO_2 films. Appl Phys Lett 26, No 8, pp 436-8, (April 1975)

238. MUGHAL H A, ECCLESTONE W and STUART R A: Spatial distribution of defects in SiO_2. Electronics Lett, 14, No 24, pp 761-2, (November 1978)

239. RAIDER S I: Time-dependent breakdown of silicon dioxide films. Appl Phys Lett 23, No 1, pp 34-6, (June 1973)

240. LI S P, BATES E T and MASERJIAN J: Time-dependent MOS breakdown. Solid-State Electron 19, pp 235-239, (1976)

241. OSBURN C M and RAIDER S I: The effect of mobile sodium ions on field enhanced dielectric breakdown in SiO_2 films on silicon. J Electrochem Soc 120, No 10, pp 1369-73, (October 1973)

242. ANOLICK E S and NELSON G R: Low field time dependent dielectric integrity. Proc 17th Ann IEEE Rel Phys Symp, pp 8-12, (1979)

243. CROOK D L: Method of determining reliability screens for time dependent dielectric breakdown. Proc 17th Ann IEEE Rel Phys Symp, pp 1-7, (1979)

244. ZALAR S M: A microphase mechanism of subintrinsic dielectric breakdown in MIS strcutures. Extd Abs Electrochem Soc Spring Mtg, pp 139-40, (May 1975)

245. BERENBAUM L: The effect of submicron particulate contamination on the properties of thin dielectric films. Extd Abs Electrochem Soc Spring Mtg 137-8, (May 1975)

246. RIDLEY B K: Mechanism of electrical breakdown in SiO_2 films. J Appl Phys 46, No 3, pp 998-1007, (March 1976)

247. OSBURN C M and CHOU N J: Accelerated dielectric breakdown of silicon dioxide films. J Electrochem Soc 120, No 10, pp 1377-84, (October 1973)

248. BARRETT C R and SMITH R C: Failure modes and reliability of dynamic RAMs: Tech Digest IEEE Electron Dev Mtg, pp 319-22, (1976)

Subject Index